THE BOAT IMPROVEMENT GUIDE

IAN NICOLSON

C. Eng FRINA Hon. MIIMS

AMBERLEY

To Alasdair Reynolds

In memory of the sea trials of *Longa* (12 feet overall)
21 January 1984, Gareloch.
Wind: Force 6 and rising.
Sea State: Rough and getting worse.
Land Conditions: Deep snow, drifting.

First published as *Comfort in the Cruising Yacht*, Nautical Books, 1986, this edition published by
Amberley Publishing, 2015.

Amberley Publishing
The Hill, Stroud
Gloucestershire, GL5 4EP

www.amberley-books.com

British Library Cataloguing in Publication Data.
A catalogue record for this book is available from the British Library.

ISBN 978 1 4456 5331 0 (print)
ISBN 978 1 4456 5332 7 (ebook)

Typeset in 10.5pt on 13pt Celeste.
Typesetting and Origination by Amberley Publishing.
Printed in the UK.

Contents

Introduction

Two of us, both young naval architects, went off for a cruise in a 12-footer. She was an entirely open boat, without even a short foredeck. Her clinker planking was aged and beginning to soften. Her sails were that grey colour that tells of years of hard work, strong winds and long days of sunshine ... all the things that make sails old and tired.

She was a pretty boat with a sweeping sheer and an old-fashioned rounded bow. We loved her dearly and would have taken her across oceans if only we had had the time. Instead, we were away for the weekend, and as dusk fell on the first night, the wind dropped completely. The calm was so perfect that the water seemed to have a glossy skin on it. We felt we could have walked to the nearby shore, across the sea's surface.

Using the last of the ebb tide, we drifted out of the main channel and stowed the limp sails when we reached the shallows. Not even a small yacht, creeping along the shore to cheat the tide, would sail so far from the channel. There were only inches below the keel when we lowered the little anchor and started brewing up coffee by the last of the fading light.

Squatting on the bottom boards, I was leaning over the single-burner primus stove, stirring the coffee and sugar into the hot milk and water. It is at moments like this that we all know why we go cruising. The silence was sharpened by the distant cry of a single curlew. Down channel we saw the stern light of the overnight ferry, which was working up speed as she reached the open sea.

I was just about to lift the pan off the stove when the wash from the ferry hit us. We were lying beam on to it. There was a brief warning as the breaking crest splashed nearby, glinting in the dusk's dim light. The next moment the boat gave a jolting roll. Too late, I grabbed the pan and cooker. There was milk, made sticky with coffee and sugar, all over the bottom boards. Mixed with it was paraffin from the primus. Some of this wet mess went on the sleeping bags we had been getting ready. The rest of it trickled through the slots in the bottom boards. We mopped up most of the mess and started brewing again.

The boat had a small persistent leak, even when moored, so we sponged her dry before going to sleep.

The first grey light of dawn was seeping through the mist when I woke. On the edge of the tide, the feeding seabirds could be heard chattering. During the night the bilge water had risen above the bottom boards, so I was lying with my cheek in a puddle of seawater, coffee and paraffin. 40 minutes later we had bailed and mopped the boat dry, cleaned up, cooked breakfast and got under way, using the dawn breeze to take us towards the open sea. There was no other sail in sight and the bliss was total.

Thirty years later my wife and I were guests on a yacht so large that her smallest boat was well over 12 feet long. We had a cabin as big as a bedroom in a city apartment. Three times a day, the talented chef produced meals that were so good we wondered if we could lure this fellow out of his galley and back to our home. Of course, we knew it was a daydream as we could not have afforded his wages for a month.

Each evening the captain, whose pay was even higher than the chef's, mixed a special brew of rum, cream and fruits and heaven knows what else, so that all activity on the ship stopped, and conversation also ceased while the first few reverent sips were savoured. Soon the rum would begin to make itself felt and the conversations would restart, slowly at first, then building to a cheerful crescendo that only hesitated when we went below to the dining saloon for another of those memorable meals.

These two wildly contrasting yachts were both comfortable, according to their owners. Comfort is not a measureable quality, mainly because it means different things to different people. It also changes according to the weather, the time of year, the size of the boat, the distance from land, the prevalence of seasickness, and the state of the drinks locker.

I've been on a boat in conditions so cold that we could not get the Calor gas to crawl out of the cylinders and down the pipes to light the stove. We had to hug the cylinders to transfer our body warmth, make the gas vapourise and flow down the pipes. At the time I remember being surprised that I did not think things were unduly tough because I had my children with me and they were enjoying themselves so much – no one could carp at any slight inconvenience. But then children in their early teens have an unlimited capacity for boisterous enthusiasm, which eliminates thoughts of discomfort.

Most people agree that the worst discomfort offshore is seasickness. The way to beat this plague is to find a pill that is effective but has no side effects, and take it in plenty of time before setting out. The drug has to be helped. It's no use dining on curry and champagne followed by sour, green apples, and expecting the pill to overcome such foods. It is best to stay on deck, but it is madness to forget to put on oilskins until the boat is out of harbour and the spray is already whipping across the deck. People prone to seasickness should make up a bunk before setting out on a voyage, even a short one, and either stay on deck or dive quickly into the bunk. It pays to undress lying down.

The diet should be of dry biscuits. Even after being sick, it is best to eat a biscuit or two. You wait perhaps 20 minutes, maybe only 10, and then start nibbling biscuits. It is important to keep the stomach supplied with something to be sick with, even if the biscuits have little chance of imparting any nutritional goodness before they return to daylight.

Next to seasickness, most people hate the cold more than anything. Cold comes most often by getting wet, so a comfortable boat is one which keeps her crew dry on deck and below. This book has a lot of ideas on this subject, and if some aspects of the matter seem to be missed, then the author's other books should be read. Having sailed in leaky boats and creaky boats, in boats with no noticeable freeboard and with hatches like colanders, I know the discomforts of being wet. Once a boat is afloat it is not so easy to stop leaks through the deck, at port holes, chain-plates and tank filler caps. In the shelter of a boatshed, or at least under a stout tarpaulin, it is not difficult to frustrate potential leaks through cockpit locker lids and forehatches. So comfort starts with good fitting out.

Once afloat, a lot of extra comfort can be harvested by being clever. When going down a channel, whether it is 10 miles wide or a 100 yards wide, it usually pays to stick close to the windward side. Here the seas will be lower, and if any sort of trouble occurs with gear or propulsion, there is sea room to leeward to give you time to reorganise.

In tidal areas, when working to windward in severe weather, the tide can often be used to ease the boat gently along towards her destination, almost hove to. Certainly progress will be slow, but it will be far more comfortable than thrashing hard into the wind. Having the engine just ticking over helps too.

Where the wind is not directly down a channel, hugging the windward side and short tacking can make the journey fun and reasonably fast, whereas out in mid-channel it will be bumpy, possibly dangerous, and highly disconcerting.

Comfort does not have awkward limitations. It does not need professional skills to make the most of the gear and gadgets shown in these pages. And even if some help is required, perhaps from a sailmaker or boatyard, for the fabrication of a part, very often you can save money by doing the fitting yourself.

Whatever is being made or altered, it pays to use the best materials, the finest fastenings, the most reliable finishing and the best components. The cost of excellent wood is little more than shoddy rubbish, and good materials often pay for themselves in time saved in making the fitting. In future years, a gadget made from good materials goes on working well long after the poorly constructed one has to be replaced. When at sea in rough conditions, if the whole crew know that everything on board is of the best, they will have peace of mind, which is the greatest comfort of all.

Ian Nicolson
Cove, 2015

CHAPTER 1

The Navigation Area

All within arm's reach of the navigator

Once settled in his seat, the navigator wants all his equipment close to hand. He does not even want to have to go across to the galley and grab a bar of chocolate, if conditions are hectic.

The layout shown here includes such conveniences as a special shelf for the next chart to be used, and a stowage slot for the charts when they have been used. This means that it is seldom necessary to lift the chart table top to rummage through the main locker for an urgently needed chart.

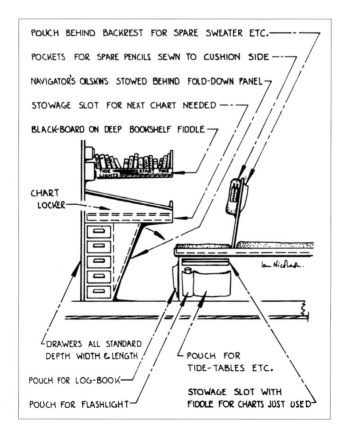

POUCH BEHIND BACKREST FOR SPARE SWEATER ETC.

POCKETS FOR SPARE PENCILS SEWN TO CUSHION SIDE

NAVIGATOR'S OILSKINS STOWED BEHIND FOLD-DOWN PANEL

STOWAGE SLOT FOR NEXT CHART NEEDED

BLACK-BOARD ON DEEP BOOKSHELF FIDDLE

CHART LOCKER

DRAWERS ALL STANDARD DEPTH WIDTH & LENGTH

POUCH FOR LOG-BOOK

POUCH FOR FLASHLIGHT

POUCH FOR TIDE-TABLES ETC.

STOWAGE SLOT WITH FIDDLE FOR CHARTS JUST USED

The front of the bookshelf is painted with that special matt black paint used on blackboards so that important information can be chalked up and quickly changed. This can be more convenient than pinning up pencilled notes.

When designing anything like a chart table it is well worth finding out what size of standard drawer is available so that the furniture can be planned round these. This is a great way to save money and time.

Tilting chart table and stool

If a yacht is pounding along on one tack for a long time, much of the discomfort of navigating can be eliminated by having a chart table that is pivoted so that it can be roughly level. If there is an attached stool for the navigator he can beaver away, well wedged in place, and all the while his paraphernalia will not slide off the chart table.

A strong base is needed to take the weight of the table and the navigator, so a set of stainless steel tubes bolted to the hull structure is called for, or a pair of bulkheads. The table is pivoted with a locking arrangement to prevent the navigator's weight from causing an upset. It is an added luxury if the electronic instruments can tilt with the table, as shown here.

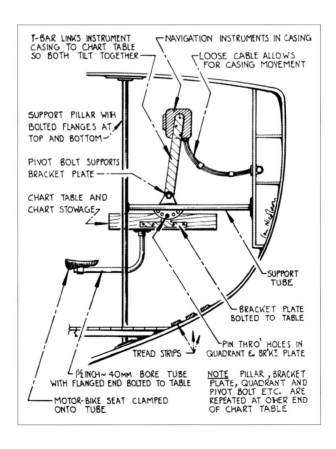

All sorts of stools can be used, but the type made from a length of tube with a tractor or motorcycle (or for small craft, a bicycle) seat clamped on is easy to assemble. The navigator splays his feet well apart and there must be tread strips to prevent him slipping.

Ideal dimensions of chart table and navigator's seat

For maximum comfort, a chair must accommodate different sizes of people by being variable. The height of the backrest, the distance from the seat front to the table front, and other critical dimensions have to be changed from person to person.

Almost as important, after anyone has been sitting in one position for a long time, relief from aches can be found by changing the tilt of the seat, sliding it back or forward, raising or lowering the footrest, and so on.

The chair and table shown here are based on researches done by V. Burandt and E. Grandjean, but modified to suit conditions in a boat. The basic seat may be a standard office chair mounted on a base, in turn set on a pair of standard seat slides as fitted in a car (but protected from corroding) or a pair of sail track lengths. The locking bolt holds the chair in place and can be an ordinary barrel bolt, or better still, a pair, one on each side.

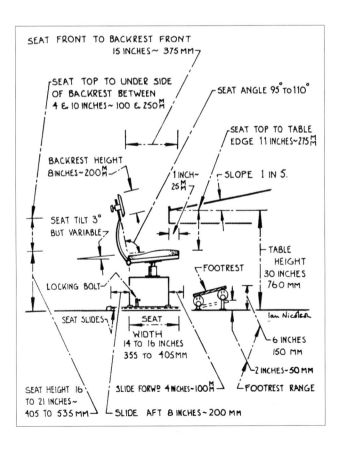

An unobtrusive chart table

The beauty of this design of chart table is that it folds into a small space when not needed, but it is not portable and is effortlessly flipped down into its in-use position in a second. Tables that are unlatched and lugged half-way across the cabin, then dropped into tricky sockets, are a menace because once at sea, no one can set them up.

The table is in two halves, which will normally be the same size, though this is not absolutely essential. The outboard edge of the outboard part is hinged to a shelf front, and the two halves are also hinged together. Some owners, especially those who are, shall we say, just a little clumsy, will want a batch of barrel bolts or turn-buttons to hold the table in the down position.

The ledges on which the table rests when down are L-shaped so that the table never scratches the bulkheads when being lowered.

A favourite material for the chart table is ½-inch (12 mm) marine ply, but the next size down can be used provided everyone is careful, and provided the inboard edge stiffener is tough. It might be 2 × 2 inches (50 mm × 50 mm) in section, glued and screwed at 5-inch (125-mm) intervals.

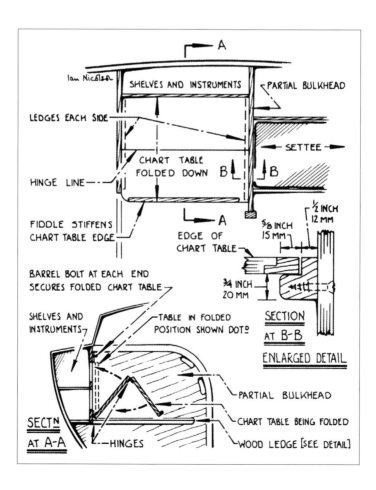

Keeping the navigator dry

In rough conditions the navigator squats on his seat at the foot of the companionway steps, trying to work out where the yacht is. Many of the electric gadgets are dead through too much moisture, taken internally. The larger waves tumble along the deck and swoosh through the main hatch, onto the navigator and all his gear. The charts become wetter than sodden toilet paper, and the navigation books soggy, with many of their pages stuck together.

To avoid this chaotic situation, a simple waterproof curtain is needed. In good weather it can be kept rolled aft, behind the navigator's seat, or rolled upwards and secured above by the hatch edge. If the latter stowage is used, the curtain will drop downwards, and may need Velcro along the sole to hold the bottom.

The curtain in the sketch unrolls and is pulled forward to the bulkhead edge where a few ties hold it in place. It can even double as a leeboard, holding the now dry and cosy navigator in his seat.

To keep the curtain from bellying, it may be necessary to have battens sewn in, as shown in the detail at the top. These battens will certainly help when stowing the curtain.

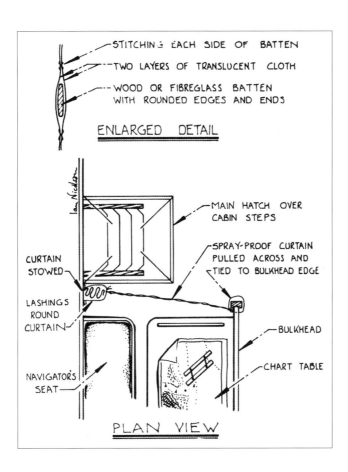

Keeping the chart dry

When the navigator goes below to work on the chart in damp conditions, his oilskins drip all over the place and water seeps or sploshes in through the hatch.

To keep the chart dry, this ¼-inch (6-mm) clear plastic cover is ideal. At the top it has one of those plastic watertight hinges; down each side it has a well-bedded fiddle, which prevents water oozing round each side; and water that collects on top drips off the bottom edge.

Ideally, the chart – which is shown dotted – is well inside the four edges of the top cover. If necessary, the chart should be folded so that it is 1 inch (25 mm) in from all four edges.

Chart work is done with that type of felt-tip pen that is easily, but not too easily, wiped off after use. The chart must be clipped immovably with drawing table springs, otherwise the marks on the cover will not be over the correct points on the chart.

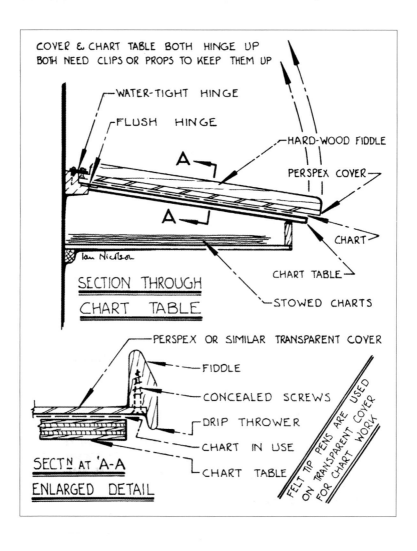

Spray deflector over the navigation instruments

In very bad weather, water may come through the main hatch and splash onto the chart table, wetting everything around. Instruments that are not waterproof suffer expensive damage. They need a cover which allows them to be used, yet protects them from waves coming aboard.

This shield consists of a Perspex front with wood top and sides. The top extends back to the bulkhead where there is a hinge which may be a broad strip of PVC material, or something similar. Secured down each side of the Perspex there are broad curtains of PVC or some other waterproof cloth. These curtains are wide enough to allow the Perspex front to be pulled away from the instruments when the navigator wants to reach up under the cover and twiddle their knobs.

To keep the shield pulled forward there may be locking friction hatch props, or 'skylight openers'. These are fittings which hold hinged covers at different angles.

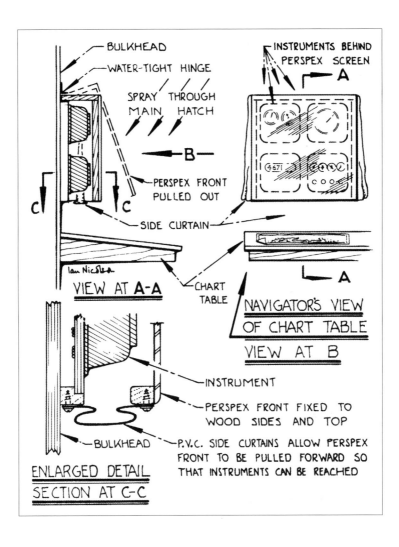

Chart table edges

The chart is indicated by the dotted line. The navigator wants an arrangement whereby his chart edge is kept flat and pinned down, yet is easy to change, even in rough conditions. Section 1 has the virtue that the chart can be seen through the Perspex, but the parallel rulers cannot operate against the clear material. Section 2 is in some ways better as the parallel rulers can be used right down to the edge of the board, and any length of chart can be fed down through the slot.

Among the virtues of Section 3 are simplicity and strength. In addition, the upstanding part of the wood edging does act as a fiddle, so if the table tilts, pencils will not roll off.

If the ends of the slot are closed there should be drain-holes at each end, otherwise in wet weather a puddle will accumulate under the lip.

Strength and stiffness are also virtues of Section 4, but as the navigator leans over the chart board he may chafe the chart, or wet it.

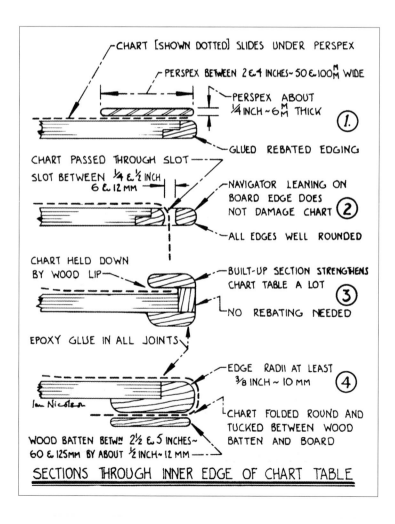

CHART [SHOWN DOTTED] SLIDES UNDER PERSPEX

PERSPEX BETWEEN 2 & 4 INCHES~ 50 & 100ᴹ WIDE

PERSPEX ABOUT ¼ INCH ~ 6ᴹ THICK   (1.)

GLUED REBATED EDGING

CHART PASSED THROUGH SLOT

SLOT BETWEEN ¼ & ½ INCH 6 & 12 MM

NAVIGATOR LEANING ON BOARD EDGE DOES NOT DAMAGE CHART   (2)

ALL EDGES WELL ROUNDED

CHART HELD DOWN BY WOOD LIP

BUILT-UP SECTION STRENGTHENS CHART TABLE A LOT   (3)

NO REBATING NEEDED

EPOXY GLUE IN ALL JOINTS

EDGE RADII AT LEAST ⅜ INCH ~ 10 MM   (4)

Ian Nicolson

CHART FOLDED ROUND AND TUCKED BETWEEN WOOD BATTEN AND BOARD

WOOD BATTEN BETWⁿ 2½ & 5 INCHES~ 60 & 125MM BY ABOUT ½ INCH~ 12 MM

SECTIONS THROUGH INNER EDGE OF CHART TABLE

Log book stowage

Unless it has a special stowage space, the log book is likely to become an orphan, or be shoved in among other books, or left sliding about on the chart table, or, perhaps, jammed in a netting rack beside the navigator's perch.

A good stowage place is a special drawer that pulls out from under the chart table. This drawer has no front, so the log book can be written up in its stowage place. To hold the book firmly as well as open at the right place, there are two strips of shock cord fixed to the bottom of the drawer at the correct distance apart to hold the pages down. These shock cords can also hold a piece of polythene sheet over the book to make doubly sure that it stays dry.

If this drawer is on the far right of the chart table it can be opened without the navigator having to move, but it will only suit a right-handed person. It could pay to have a drawer on each side of the chart table, and use the second for the nautical almanac or the tide tables.

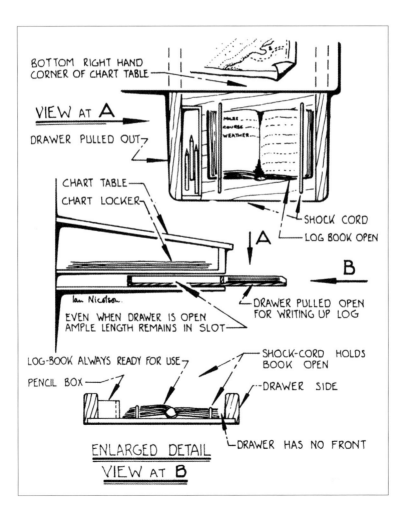

A pencil (and everything else) holder

The usual type of pencil holder, positioned over chart tables the world over, has an upper and lower shelf. The top shelf has a row of holes drilled in it, each hole slightly larger than the diameter of a pencil. Only pencils fit through these holes, but navigators and the rest of the crew need a 'universal' holder, which keeps all sorts of other things safe and ready to hand.

This 'universal' holder has two thick strips of that squashy foam plastic used inside modern cushions, with a gap of about ⅛ inch (3 mm) between the two strips. All sorts of small gadgets can be pushed down between the twin cushions and they will still be there after sailing through a gale.

The length of this holder can be as long as the chart table is wide, but in practice this will mean that everything from the bread knife to the skipper's false teeth will end up there. So it is probably best to make this rack about 10 inches (250 mm) long.

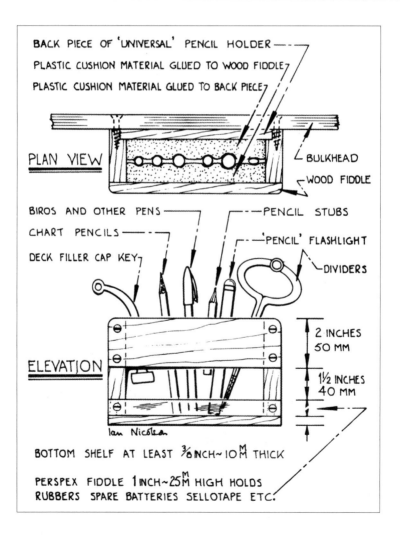

Navigator's noticeboard

It is so handy to have a display area over the chart table, where important information can be arrayed. Tide times for the weekend, a reminder to replenish the gas bottles, train times for getting home. These are the sorts of information we need near at hand.

However, not everyone likes having drawing pins on a boat. They rust and they fall on the sole, sharp point upwards, which is no fun for barefoot crew.

A clothes peg (or clothes pin as some people prefer to say) board has lots of advantages. It does away with drawing pins, and the pegs can be colour-coded so that important information is held up by the brightest red pegs, and less important items by duller pegs. Each peg can hold several sheets of paper, or even things like the class racing flag, a spare pencil, or cash for the harbour master's dues.

A single row of pegs suits most boats, but some people prefer an upper and lower row, so as to retain the top and bottom edges of each piece of paper.

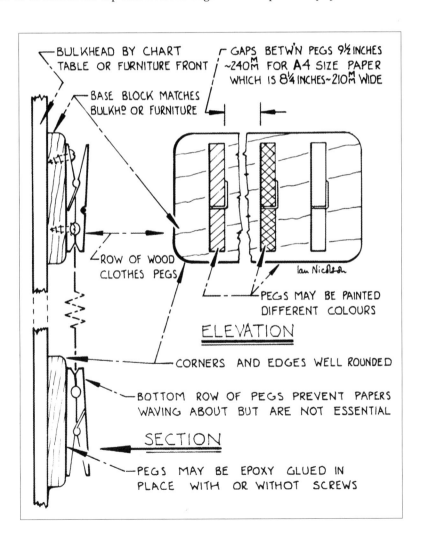

CHAPTER 2

Galleys

Warm plates and gimballed cookers

A meal is so much better if the plates are warmed, and this can be done painlessly if the plate storage shelf is made of stainless steel and is located above the cooker. There should be large holes in the shelf, and it must be small enough to keep the plates from slamming about when the boat rolls.

The gimballing of cookers is an art. For a start, the pivot level must be high but not too high, and the ballast, in the form of the stove and the pans on it, must be heavy but too heavy. Experimentation is the best way to find the right location for the pivot and the correct weight. In general, the pivot should be high, perhaps level with

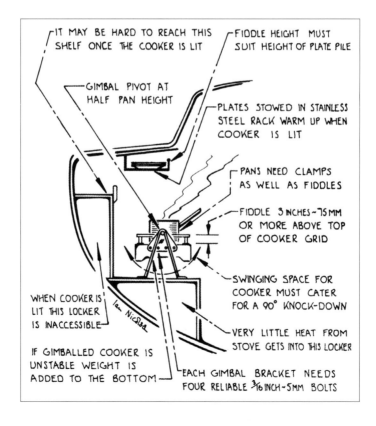

IT MAY BE HARD TO REACH THIS
SHELF ONCE THE COOKER IS LIT

FIDDLE HEIGHT MUST
SUIT HEIGHT OF PLATE PILE

GIMBAL PIVOT AT
HALF PAN HEIGHT

PLATES STOWED IN STAINLESS
STEEL RACK WARM UP WHEN
COOKER IS LIT

PANS NEED CLAMPS
AS WELL AS FIDDLES

FIDDLE 3 INCHES ~ 75 MM
OR MORE ABOVE TOP
OF COOKER GRID

SWINGING SPACE FOR
COOKER MUST CATER
FOR A 90° KNOCK-DOWN

WHEN COOKER IS
LIT THIS LOCKER
IS INACCESSIBLE

VERY LITTLE HEAT FROM
STOVE GETS INTO THIS LOCKER

IF GIMBALLED COOKER IS
UNSTABLE WEIGHT IS
ADDED TO THE BOTTOM

EACH GIMBAL BRACKET NEEDS
FOUR RELIABLE 3/16 INCH ~ 5MM BOLTS

the middle of the pans unless there is an oven, in which case it can be one third of the way up the deepest pan. Weight can always be added to the bottom of a cooker to make it more stable.

Standard cooker fiddles are nearly always too low; they should be at least 3 inches (75 mm) high, even if they have to have cut-outs for the pan handles.

Plan view of the 'feeling 1040' galley

The cook works safely in the confines of the C-shaped galley; in bad weather there is little chance of being thrown far. On the right of the sink is the saltwater pump, so it is the one most likely to be used as most people are right-handed, and this tends to save fresh water!

The secondary sink on the right can be used to hold dirty dishes until there is time to wash them, or, after washing, they can be left to drain in this mini-sink, which is narrow but also deep and safe. Ahead of it is a deep locker for stowing three large bottles. The deckhead support, in the form of the stainless steel pillar, is oval in section, so as to take up the minimum space athwartships.

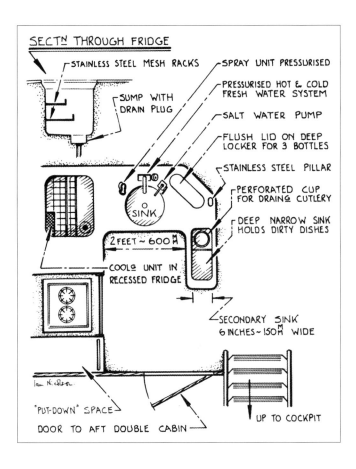

In the fridge there are two shelves, each with fiddles, set in such a way that the top one is narrower than the lower one, to give a view of the food stowed on the 'lower deck'. The sump makes it easy to dry out the fridge after defrosting, and washing out is helped by the presence of a proper drain complete with plug.

Twin sinks and their surrounds

Sinks should not be fitted hard up against a bulkhead, otherwise the person washing up lacks elbow room. If the space between sink and bulkhead is quite small, and it needs to be no more than 6 inches (150 mm), it is too narrow for putting down plates. But it can be made into a trough-type draining board. This consists of a watertight recessed box made of metal, wood or fibreglass, into which wet cups, plates and cutlery can be put.

That narrow space beyond the sinks is ideal for another trough, less deep, and designed to hold standard plastic or glass containers for foods like salt, sugar, tea and so on. This trough should be designed so that the containers drop in easily.

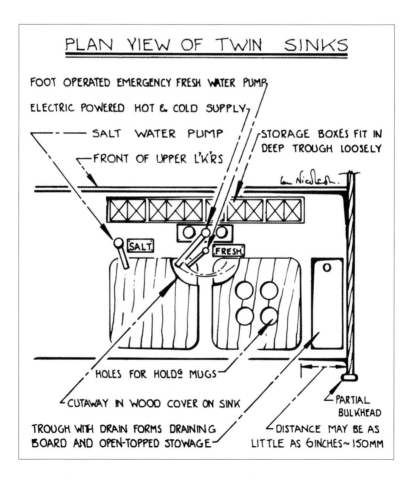

PLAN VIEW OF TWIN SINKS

FOOT OPERATED EMERGENCY FRESH WATER PUMP

ELECTRIC POWERED HOT & COLD SUPPLY

SALT WATER PUMP

FRONT OF UPPER L'K'RS

STORAGE BOXES FIT IN DEEP TROUGH LOOSELY

SALT

FRESH

HOLES FOR HOLD^G MUGS

CUTAWAY IN WOOD COVER ON SINK

TROUGH WITH DRAIN FORMS DRAINING BOARD AND OPEN-TOPPED STOWAGE

PARTIAL BULKHEAD

DISTANCE MAY BE AS LITTLE AS 6 INCHES ~ 150MM

The wood covers on the sinks form cutting boards; they have one corner cut away to take the drips from the pumps, and to allow them to be easily lifted off. One board is cut to take mugs when they are being filled.

More working space for the cook

To give the cook more 'put-down' space, just make up a big tray and screw it down on top of the existing working surface. The tray is made larger than the existing worktop by extending it inboard, forward, and possibly aft. The forward overlap steals a little space over the settee, but does not stop three or four people sitting there. The inboard extension is acceptable because when people stand close to a work surface, they need more room at sole level than at worktop level.

The neatest way to make the new tray is to laminate the fiddle, and for seagoing use this should be good and high. The enlarged detail shows how to bend the laminates round a curved tin to get lovely corners. It is usual to have a hard-wearing plastic laminate on the worktop surface, but some people prefer to use an epoxy resin paint or varnish.

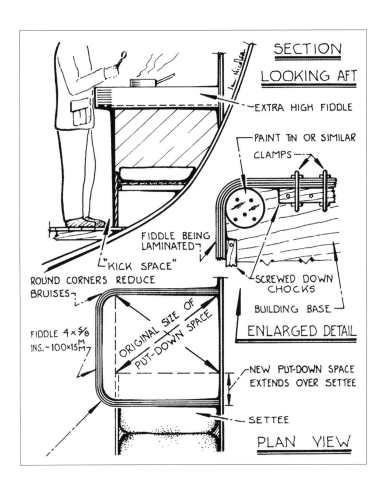

Gimbals for a big cooker

A set of gimbals made to this design carries the weight of a heavy cooker. The parts can be made away from the cooker because sizes do not have to be accurate to the nearest ¼ inch (6 mm), provided the base width is slightly greater than the cooker width.

The side straps bend easily in at the top and middle to hug the cooker, and the fastenings through them can be located wherever it suits. The channel bars are drilled with rows of holes so that the pivot height can be varied. Typically, the channels will be 8 inches (200 mm) from top to bottom. The bottom flat bars under the cooker will be separated by a distance of about three-quarters the athwartships dimension of the cooker, unless they simply must be set under the cooker feet.

A barrel bolt or other locking device is essential to stop the cooker swinging when the gimballing effect is not needed.

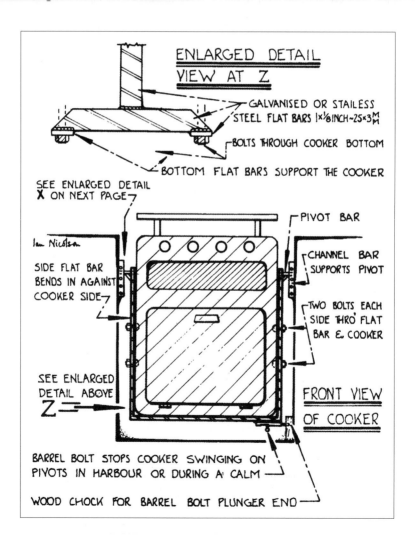

ENLARGED DETAIL VIEW AT Z

GALVANISED OR STAILESS STEEL FLAT BARS 1x⅛INCH~25×3 M/M

BOLTS THROUGH COOKER BOTTOM

BOTTOM FLAT BARS SUPPORT THE COOKER

SEE ENLARGED DETAIL X ON NEXT PAGE

PIVOT BAR

Ian Nicolson

SIDE FLAT BAR BENDS IN AGAINST COOKER SIDE

CHANNEL BAR SUPPORTS PIVOT

TWO BOLTS EACH SIDE THRO' FLAT BAR & COOKER

SEE ENLARGED DETAIL ABOVE Z

FRONT VIEW OF COOKER

BARREL BOLT STOPS COOKER SWINGING ON PIVOTS IN HARBOUR OR DURING A CALM

WOOD CHOCK FOR BARREL BOLT PLUNGER END

Gimbal details

At the top is a sketch showing an alternative way of supporting the cooker. Instead of horizontal flat bars, a pair of angle bars take the weight of the cooker. They fit under the cooker feet with bolts through the feet.

The bottom two sketches are doubly enlarged to show the details of the pivoting. The pivot bars have triangular brackets under, and rest on rubber blocks. These dampen the swing. If the boat inverts, the locking bolts above the pivots prevent the cooker romping away from the galley and terrorising the crew.

There must be three bolts through each channel bar of about $\frac{3}{16}$-inch (5-mm) diameter, and they need not be countersunk provided they are clear of the pivot bars.

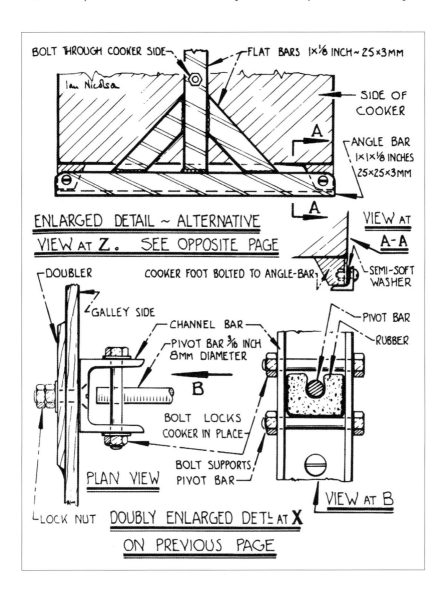

Gimballed working surface

When far out at sea and slaving over the stove, there is nothing so handy for the cook as a level surface for putting down cups and plates. When soup has to be taken off the stove and poured into mugs, the job is so much easier, and if a stew has to be spooned into plates, life is less fraught, if the plates are on a gimballed surface.

This design makes use of the gimballed cooker to which it is attached. Three pivot points are needed and they must be exactly in line. All the ballast is provided by the cooker, so it may be necessary to bolt a steel plate to the bottom to counteract the weight of a fully laden pan resting on the high gimballed put-down space.

High fiddles are essential at sea because plates and mugs can slide, even on a gimballed surface.

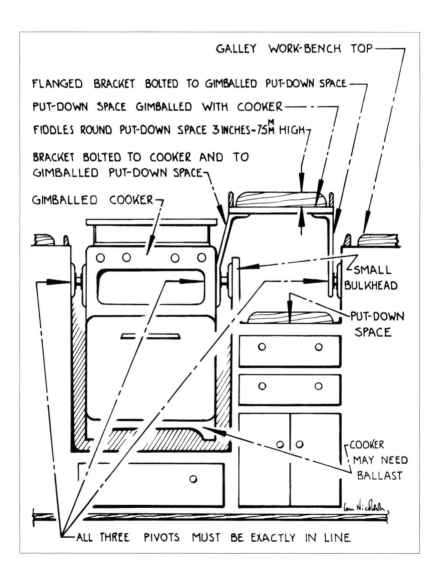

GALLEY WORK-BENCH TOP

FLANGED BRACKET BOLTED TO GIMBALLED PUT-DOWN SPACE

PUT-DOWN SPACE GIMBALLED WITH COOKER

FIDDLES ROUND PUT-DOWN SPACE 3 INCHES ~ 75 M HIGH

BRACKET BOLTED TO COOKER AND TO GIMBALLED PUT-DOWN SPACE

GIMBALLED COOKER

SMALL BULKHEAD

PUT-DOWN SPACE

COOKER MAY NEED BALLAST

ALL THREE PIVOTS MUST BE EXACTLY IN LINE

Converting the chart table for galley use

Few galleys have enough working space, and many cooks end up with plates and cups on the cabin sole. It so happens that on many boats the chart table is opposite the galley, and in harbour the chart table is seldom in use. So, if it can be used by the cook, so much the better.

A sloping chart table is easily levelled with simple but strong turn-buttons, one at each side. A single turn-button is not a good idea because at times the cook may use the chart table for cutting up tough meat, or slicing through hard cheese, and then the load will be heavy.

Some details in the sketch concern the navigator. So that the charts lie absolutely flat on the table, the hinges should be the flush type, with no part standing up from the smooth surface. The best place for the pencil shelf is high at the back, because this uses space otherwise wasted. That hole in the chart locker bottom is a real asset not only when dusting out, but also for pushing up a heavy heap of charts so as to retrieve one buried in the pile.

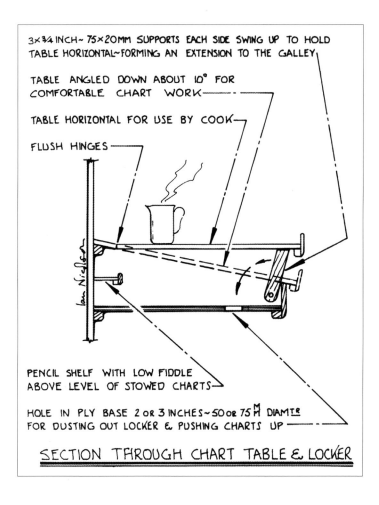

3×¾ INCH~ 75×20mm SUPPORTS EACH SIDE SWING UP TO HOLD TABLE HORIZONTAL~FORMING AN EXTENSION TO THE GALLEY

TABLE ANGLED DOWN ABOUT 10° FOR COMFORTABLE CHART WORK

TABLE HORIZONTAL FOR USE BY COOK

FLUSH HINGES

PENCIL SHELF WITH LOW FIDDLE ABOVE LEVEL OF STOWED CHARTS

HOLE IN PLY BASE 2 OR 3 INCHES~50 OR 75mm DIAMtr FOR DUSTING OUT LOCKER & PUSHING CHARTS UP

SECTION THROUGH CHART TABLE & LOCKER

Mug rack for all weathers

If a yacht is slamming into a head sea, or rolling horribly in a swell, the crew will find a simple rack for mugs near the cooker, a real boon. This rack will suit a variety of galleys, and can be easily made by amateur or professional. The ideal material is a wood that matches the rest of the trim round the galley.

The rack can be fixed permanently, or, as in the sketch, made portable. It will normally be in place all the time the boat is at sea, and the cook can fill the mugs without taking them from the rack.

A similar rack could be made to take deep soup bowls.

If the top of the partial bulkhead is horizontal, and the top moulding is shallow, the rack will tend to tip sideways, at least slightly, so it will be necessary to glue chocks to the bulkhead to fit the 'skirts'. These chocks will thicken the bulkhead locally and ensure that the inside faces of the skirts are held rigidly.

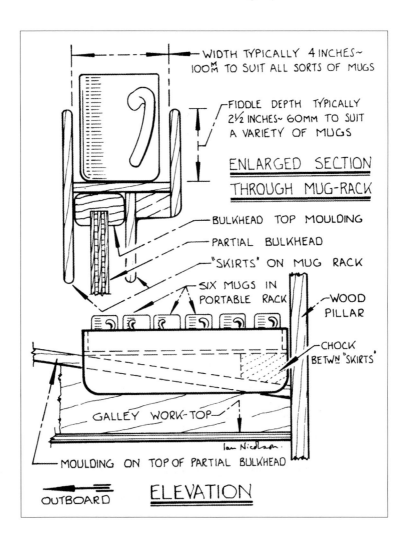

A drawer beneath the sink

Because sinks have to have drains, the space beneath the sink bowl is often wasted. However, a clever joiner can make up a special drawer, shaped like a thick U, to slide under the sink and yet avoid all the piping.

The base of the drawer should be of marine ply, probably about ⅜ inch (10 mm) thick, to give the drawer ample rigidity. The ends will be of hardwood, typically ¾ inch (20 mm) thick, but the sides can be lighter. If the whole drawer is assembled and glued up with epoxy resin, it will gain the stiffness it needs to cope with its odd shape.

Anyone who has a drawer, or indeed any bit of furniture, which is becoming floppy can stiffen it by gluing in fillets at the corners and edges.

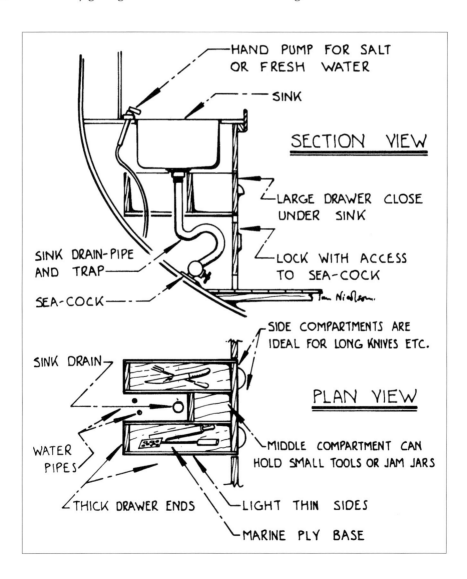

Frying pan locker

A frying pan takes up a lot of space, and because of its shape, it is not easy to stow in such a way that it stays where it is put. This special locker uses up what is sometimes a wasted space beside the cooker, or maybe between the sink and the edge of the galley bench.

The special locker is narrow, and just high enough to accept the pan through the slot in the front. The pan drops down inside the locker and jams automatically between the locker front and the soft foam plastic cushioning at the back. This cushion may be covered with leather cloth or a similar washable material.

To clean out the locker, remove the pan, turn the top and bottom turn-buttons and remove the front panel. Ideally, the inside should be lined with a smooth, washable plastic material like that used on galley bench tops.

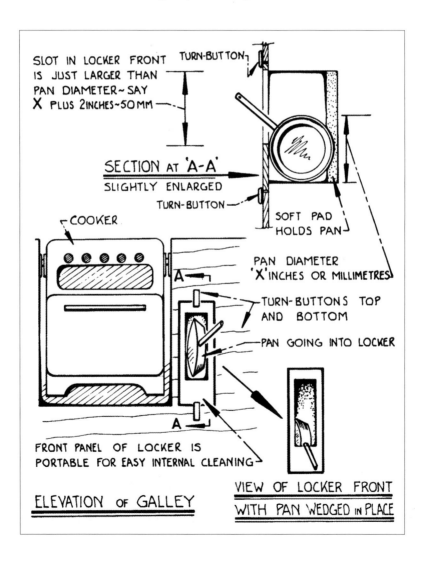

Cook's comfort

A strong pipe, bent like an outsize croquet hoop and bolted to the sole in the galley, makes a great prop for the cook. On one tack the cook leans against the hoop, and on the other the 'galley strap' is hooked onto each side of the hoop to hold the cook off the stove. For extra comfort there is a wooden pad for the cook to lean on, and this can have fiddles to hold the cook in place if the boat is pitching a lot. These fiddles are not shown in the drawing, but are at each end of the pad.

The hoop may be made to fit one particular person, as shown top left, or the crossbar can be varied up and down to suit a range of heights, as detailed top right. The hoop may extend the full length of the galley and could even have two pads, though not every cook likes an assistant working alongside.

The base of each vertical tube must be strongly bolted, never screwed, to the cabin sole, with suitable doublers to take the shock loadings of people lurching against the bars.

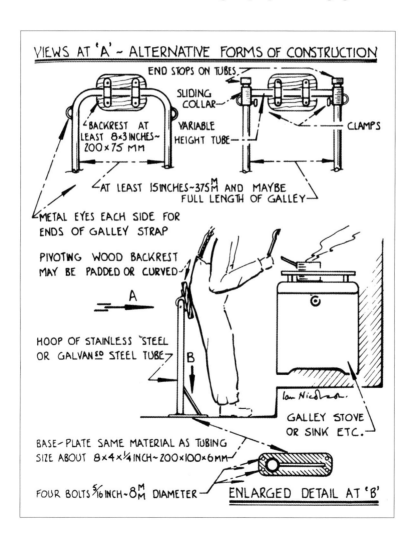

VIEWS AT 'A' - ALTERNATIVE FORMS OF CONSTRUCTION

END STOPS ON TUBES

SLIDING COLLAR

BACKREST AT LEAST 8×3 INCHES ~ 200×75 MM

VARIABLE HEIGHT TUBE

CLAMPS

AT LEAST 15 INCHES ~ 375 M AND MAYBE FULL LENGTH OF GALLEY

METAL EYES EACH SIDE FOR ENDS OF GALLEY STRAP

PIVOTING WOOD BACKREST MAY BE PADDED OR CURVED

A

HOOP OF STAINLESS STEEL OR GALVANSD STEEL TUBE

B

GALLEY STOVE OR SINK ETC.

BASE-PLATE SAME MATERIAL AS TUBING SIZE ABOUT 8×4×¼ INCH ~ 200×100×6MM

FOUR BOLTS 5/16 INCH ~ 8 M DIAMETER

ENLARGED DETAIL AT 'B'

Dealing with rubbish

Call it a gash bin, a trash bucket, a rubbish box or a garbage container. Call it what you like. It holds all the waste material thrown out by the cook during the preparation of a meal. This design fits the available space under the galley worktop even if the boat slopes in sharply, or it can be squeezed into the corner of an engine bay, or under a cooker ... in short, it can be worked into all sorts of odd-shaped spaces.

The basic box will probably be made of ½-inch (12-mm) marine ply, or on a racing boat out of ⅛-inch (3-mm) marine ply with lightening holes and stiffeners. It will not become particularly dirty as it is lined with the same sort of plastic bags that are sold for kitchen use. To prevent the bin swinging open when the boat heels, there is a simple pivoted bar, made from a length of brass about 8 inches × ¾ inch × ⅛ inch (200 mm × 20 mm × 3 mm). This locking device can be used on other types of stowage, and the whole bin concept can be used for keeping things other than kitchen waste.

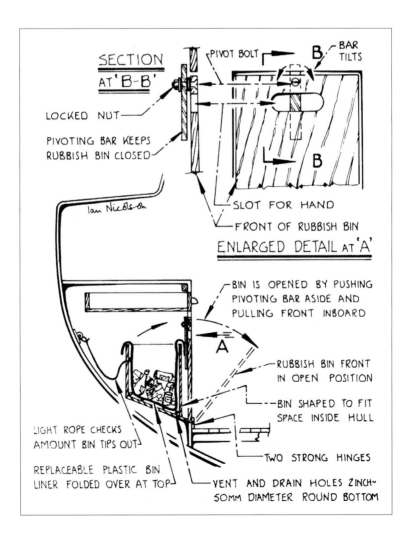

SECTION AT 'B-B'

PIVOT BOLT

B — BAR TILTS

LOCKED NUT

PIVOTING BAR KEEPS RUBBISH BIN CLOSED

Ian Nicolson

B

SLOT FOR HAND

FRONT OF RUBBISH BIN

ENLARGED DETAIL AT 'A'

BIN IS OPENED BY PUSHING PIVOTING BAR ASIDE AND PULLING FRONT INBOARD

A

RUBBISH BIN FRONT IN OPEN POSITION

BIN SHAPED TO FIT SPACE INSIDE HULL

LIGHT ROPE CHECKS AMOUNT BIN TIPS OUT

TWO STRONG HINGES

REPLACEABLE PLASTIC BIN LINER FOLDED OVER AT TOP

VENT AND DRAIN HOLES 2 INCH~ 50MM DIAMETER ROUND BOTTOM

More rubbish

When cooking, it is amazing how much rubbish is produced. There are potato peelings, empty tins, plastic wrappings and much more. All this has to be put into a container with minimum fuss and inconvenience. A very good arrangement is shown here. It consists of a bin or bucket located below the galley worktop, with a hole in the worktop through which all the rubbish is pushed. The hole is covered by a simple lid that has a 'lift-up' handle let into it.

The hole in the worktop is made just big enough to take a large tin or jam jar. A larger aperture would be obtrusive. The container below the hole can be a bucket or plastic bin or a specially made box of wood, metal and fibreglass. Its size should suit the standard plastic bag designed for domestic rubbish bins, that is, about 27 inches (675 mm) deep, and the perimeter of the top rim must be a little under 46 inches (1200 mm) so that the plastic bag can be folded over it.

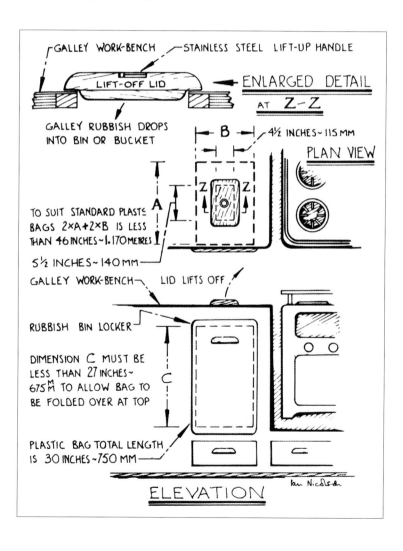

A galley strap

The bottom drawing shows the type of comfortable galley strap that is made from a length of Terylene webbing and a broad pad of sailcloth sewn to the middle of the belt. This wide backrest prevents the webbing from digging into the cook's back, and sometimes has pockets for a cooking knife, or cloth or tin opener.

The middle two sketches show a standard adjustable buckle, which makes it easy for the cook to alter the length of the belt with one hand. There must be four $^{3}/_{16}$-inch (5-mm) bolts through the baseplate of the buckle and the same number through the eyeplate which takes the snap hook on the other end of the belt. Those light alloy carbine hooks favoured by climbers make good clips for the belt end.

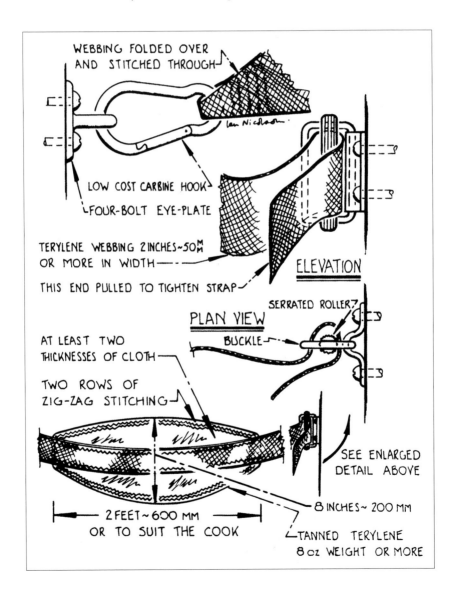

WEBBING FOLDED OVER AND STITCHED THROUGH

Ian Nicholson.

LOW COST CARBINE HOOK

FOUR-BOLT EYE-PLATE

TERYLENE WEBBING 2 INCHES ~ 50 M/M OR MORE IN WIDTH

THIS END PULLED TO TIGHTEN STRAP

ELEVATION

PLAN VIEW

SERRATED ROLLER?

BUCKLE

AT LEAST TWO THICKNESSES OF CLOTH

TWO ROWS OF ZIG-ZAG STITCHING

SEE ENLARGED DETAIL ABOVE

2 FEET ~ 600 MM OR TO SUIT THE COOK

8 INCHES ~ 200 MM

TANNED TERYLENE 8 oz WEIGHT OR MORE

Cook's safety strap

When working in the galley in rough sea conditions, the cook needs a strap to prevent him, or her, being flung about. As cooks vary in height, some arrangement is needed to make it easy to shift the strap up and down. On the left there is a length of track such as is normally bolted down on deck to take genoa sheet leads. It can be the lightest available track, often 1 inch (25 mm) wide. A standard slider, of the type normally used to take the shackle that holds the sheet lead block, is fitted on the track. A spring-loaded plunger holds the slider at the correct height, and the snap shackle on the safety strap engages in the metal eye on the slider. At the other end a lashing can be used, and this will allow the cook to vary the tension of the strap.

On the right there is a vertical handrail, with closely spaced slots. These slots must be wide enough for a large man's hand, so this means the 'jump' from one strap location to the next must be 4½ inches (110 mm), which may not suit everyone.

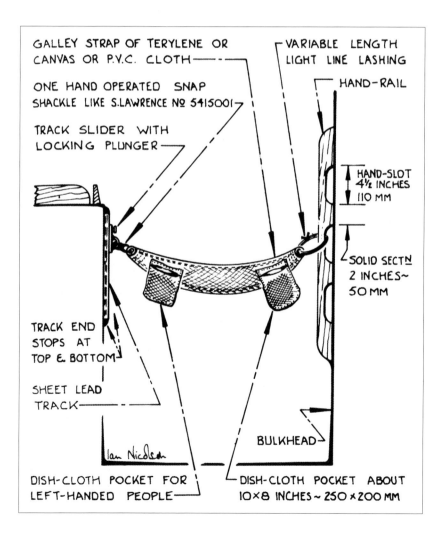

GALLEY STRAP OF TERYLENE OR CANVAS OR P.V.C. CLOTH

VARIABLE LENGTH LIGHT LINE LASHING

ONE HAND OPERATED SNAP SHACKLE LIKE S.LAWRENCE № 5415001

HAND-RAIL

TRACK SLIDER WITH LOCKING PLUNGER

HAND-SLOT 4½ INCHES 110 MM

SOLID SECT<u>N</u> 2 INCHES ~ 50 MM

TRACK END STOPS AT TOP & BOTTOM

SHEET LEAD TRACK

Ian Nicolson

BULKHEAD

DISH-CLOTH POCKET FOR LEFT-HANDED PEOPLE

DISH-CLOTH POCKET ABOUT 10×8 INCHES ~ 250 × 200 MM

Toast for breakfast

Some cookers have no grills for making toast, some have grills that make dreadful toast and some have grills that take hours to make a couple of pieces. This simple gadget makes good toast at a reasonable speed, and as it is not too bulky to stow, there is no reason why the average cruiser should not have one for each top burner on the cooker.

The bread may need moving about slightly to give it that even brown crispness we all so much enjoy, and it will probably be advisable to have a pair of tongs to turn the toast over because every part of this toaster gets hot.

The height from the top of the cooker's burner to the underside of the lowest layer of wire mesh is important, so it will be best to make this about 1 inch (25 mm) at first and gradually shorten it to about ¾ inch (20 mm) to allow fast toasting with no burning.

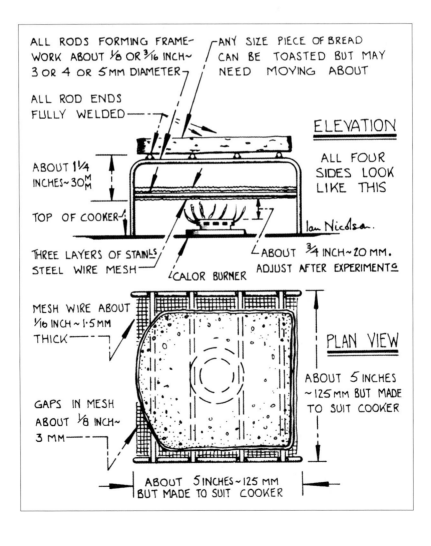

CHAPTER 3

Fridges and Iceboxes

Icebox or fridge

This design for an icebox can be used for making a fridge, if a bought-in cooling unit is used. Though mass-produced fridges are cheap, they are not always easy to install. They seldom make the best use of available space, they are rectangular, which does not suit the average boat and they are seldom really well insulated.

This drawing shows a good type of fridge or icebox with a simple type of lid that does not have to be fitted accurately. It is a bonus that this lid doubles as a chopping board for cooking.

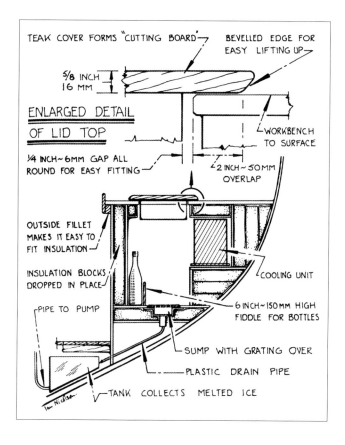

Melted ice drains into a tank where it can be pumped off, a better practice than the more usual arrangement whereby the water drains into the bilge. Though this fridge or icebox is shown under the galley bench, it can be situated under a sideboard or even under a chart table, provided the navigator is able to sit adjacent in comfort.

Icebox design

By using a 'stepped section' for the interior of this icebox, extra space is gained. The top shelf can be used for a cooling unit run off a battery, turning the icebox into a fridge.

The interior casing is made of fibreglass or ply lined with fibreglass. It is carefully placed in position and the gaps between it, the hull, the bottom plinth and the side walls are filled with foam polymer, which is poured in. Before pouring the foam the drain is fitted, and all likely leak holes where the polymer might escape are sealed off.

Most cooks like a smooth, uninterrupted worktop so the cold box lid is made flush fitting, with recessed hinges and lifting ring. The lid is bound to be at least slightly bulky and heavy, so it is not a good idea to omit the catch that holds the lid, or make the lid portable. If the lid is not a tight fit, heat will seep in and reduce the effectiveness of the cool box.

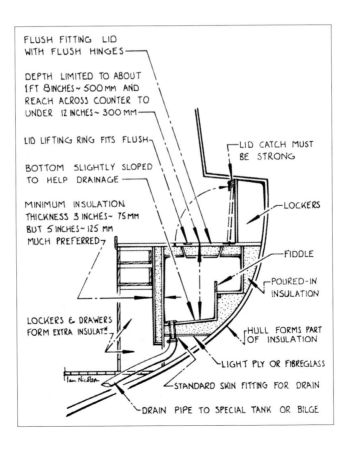

Icebox details

Only marine-grade plywood should be used when making an icebox, because there is
plenty of moisture about and lesser grades of wood will rot and disintegrate too soon.
The thicker the ply the better the insulation – wood is remarkably good at keeping
the heat out.

The drain fitting is a plastic skin fitting sold in chandlery shops. The type with
a thin flange is best if good drainage is required. One leaf of the lid hinge is shown
bolted through, and the other held with long screws into a doubler that extends round
the opening. To make the glassing in of the icebox straightforward, the heaviest cloth
used should be 1½ oz chopped strand mat, and even then all exterior edges should
be well rounded, while interior corners will need filleting. The fillets can double as
joining strips for the sides, top and bottom, but both glue and screws are needed at
all wood joins.

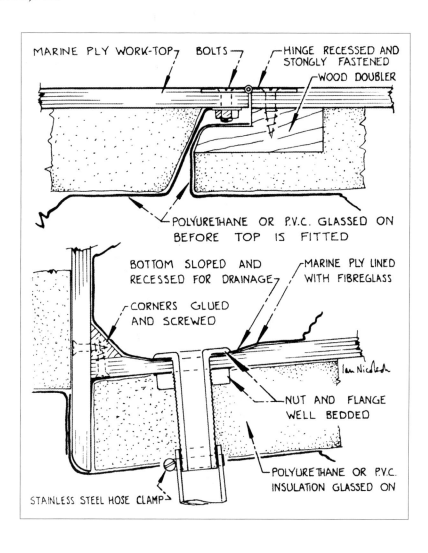

A built-in icebox or fridge

To make a container for keeping food and drinks cold, a well-insulated box is needed. This can be made of ply, with blocks of insulating material secured to the outside by simple glassing-on techniques. The sides of the box in this sketch reach well below the bottom to take extended side insulation.

A plinth is made to support the icebox, and the box itself is designed to make the best use of available space. Before the box has been lined with fibreglass it is lowered in behind the galley front, and the galley worktop (which has been previously assembled with its insulation) is fixed down. The interior of the icebox is now glassed in, though in practice one or two layers of the glass may have been applied before fitting the top.

For the drainpipe only plastic materials are used, because metal will act as a conductor and reduce the effectiveness of the insulation.

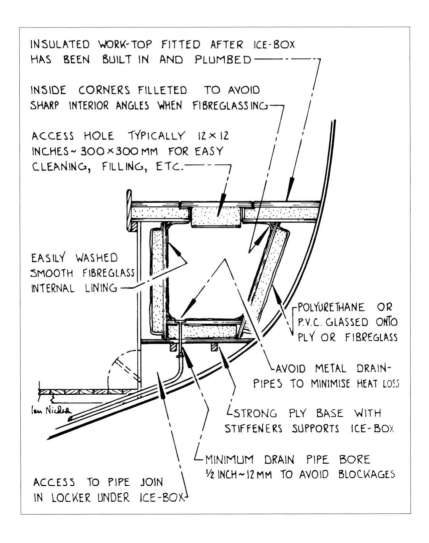

INSULATED WORK-TOP FITTED AFTER ICE-BOX HAS BEEN BUILT IN AND PLUMBED

INSIDE CORNERS FILLETED TO AVOID SHARP INTERIOR ANGLES WHEN FIBREGLASSING

ACCESS HOLE TYPICALLY 12 × 12 INCHES ~ 300 × 300 MM FOR EASY CLEANING, FILLING, ETC.

EASILY WASHED SMOOTH FIBREGLASS INTERNAL LINING

POLYURETHANE OR P.V.C. GLASSED ONTO PLY OR FIBREGLASS

AVOID METAL DRAIN-PIPES TO MINIMISE HEAT LOSS

STRONG PLY BASE WITH STIFFENERS SUPPORTS ICE-BOX

MINIMUM DRAIN PIPE BORE ½ INCH ~ 12 MM TO AVOID BLOCKAGES

Ian Nicolson

ACCESS TO PIPE JOIN IN LOCKER UNDER ICE-BOX

An under settee icebox or fridge

A long, fairly shallow icebox can be fitted under a settee, provided the crew do not mind the inconvenience of lifting the cushion to get at the food. The cushion improves the insulation, as does the dead air space round the box. By making the box long, it can either have a very good capacity, or one end can hold the refrigerating unit, if the old-fashioned icebox is not acceptable. Electrically driven cooling units can be bought from chandlers and bolted in place inside the insulated container.

The handle for lifting out the lid must be recessed, and the surface flange all round must be made quite thin, otherwise the cushion will not lie flat on the settee base. Where the flange meets the box there has to be a full-rounded fillet for strength. So that the fibreglass does not chip, the lid and box edges are made with soft radii, as shown in the sketch.

When cleaning, painting or repairing the hull, the top flange screws are removed and the whole casing is lifted out.

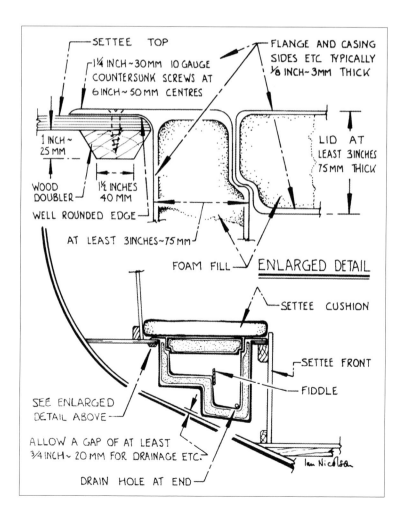

A low-price icebox

The quickest and cheapest way to fit an icebox in any boat is to buy one of the portable boxes sold by shops catering for campers and caravanners. These boxes are available in a variety of sizes and shapes, though most are longish and have about the same height as width.

Because an icebox is obtrusive in most cabins, the first thing to do is to investigate suitable stowage places and measure up each one. Then, a selection of different boxes are measured in shops to see which one will fit. There must be room to lift the box in place, and spare at the sides for at least two lashings. The crew will want to get at the food without moving a lot of furniture and certainly without moving the box from its stowage space.

Shown here are some of the possible places where an icebox might be stowed. If the worst comes to the worst, it can be banished to a cockpit locker, or to the fo'c'sle, but these are less convenient places than near the galley.

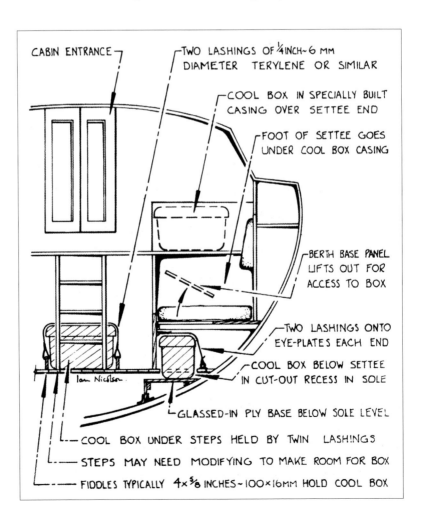

Cool-box stowage

A bulky cool-box may not fit in a cabin and that odd-shaped space inside the counter could be just the place for it. If the access hatch in the deck is too small, don't make it bigger: it's much better to saw out a panel at the aft end of the cockpit well, to get the box in place.

It is important that the well side is made truly watertight, which explains why the sealing strip is wide and the fastenings close together. When the panel is out, it will be easy to glass-in the two rests for the box and the high end fiddles that are so important. No one wants lashings over the top of a cool-box if they can be avoided, as they have to be taken off every time the lid is lifted.

The techniques shown here can be used for stowing other bulky items, such as generators, extra fuel tanks that are piped up, and perhaps extra batteries.

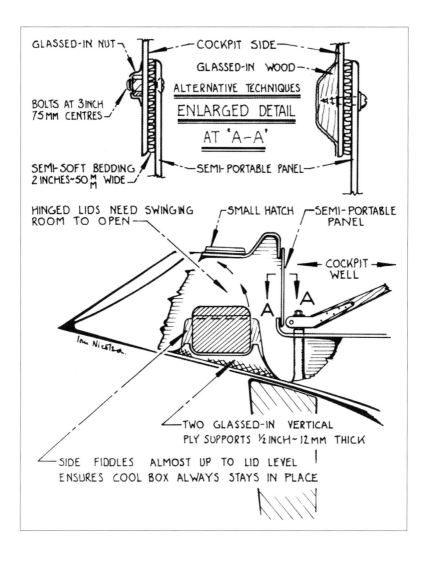

A quick, cheap icebox

This sketch shows what must be the quickest way of fitting a boat with an icebox. A standard mass-produced box that will fit in the aft end of the cockpit is bought and secured in place with a fiddle across the fore end. One or two side fiddles may be needed too, or maybe just a chock of wood between the icebox and the cockpit well sides will be adequate.

To keep the icebox in place in all conditions, a heavy cloth cover is bolted down across the fore end of the aft deck, and stretched over the box. A tight lacing or row of thick shock cord loops keeps the cover down. If the cover is white it will not get hot even in bright sunlight, and if it is of double or triple thickness it will not only be extra strong, but it will also provide better insulation. Even if the box is not as wide as the cockpit well the cover should be, so that the sun cannot reach the sides of the icebox.

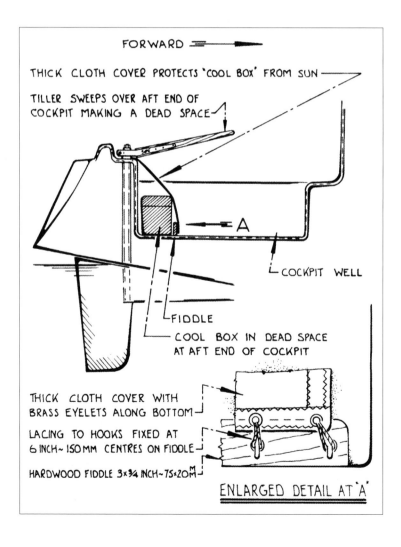

FORWARD

THICK CLOTH COVER PROTECTS 'COOL BOX' FROM SUN

TILLER SWEEPS OVER AFT END OF COCKPIT MAKING A DEAD SPACE

A

COCKPIT WELL

FIDDLE

COOL BOX IN DEAD SPACE AT AFT END OF COCKPIT

THICK CLOTH COVER WITH BRASS EYELETS ALONG BOTTOM

LACING TO HOOKS FIXED AT 6 INCH ~ 150 MM CENTRES ON FIDDLE

HARDWOOD FIDDLE 3 x ¾ INCH ~ 75 x 20 M

ENLARGED DETAIL AT 'A'

CHAPTER 4

Leeboards, Furniture and Handrails

Lee-cloths

Berth leeboards, which are made from a cloth material, are sometimes referred to as lee-cloths. They may be made from a Terylene (Dacron) cloth available at sailmakers, or an acrylic material, such as is sometimes used for sail covers and cockpit tents. If a dark material is used it will show the dirt less.

Top and bottom securing arrangements must be strong. For instance, those brass eyelets that are hammered in are not strong enough for the upper lashings, so either

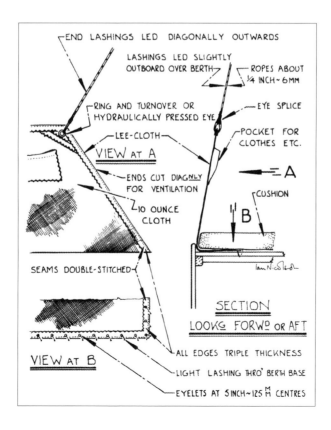

the traditional 'ring-and-turnover' or the modern equivalent, the hydraulically pressed eye, must be worked in.

It is common practice to screw a wood batten over the edge of a lee-cloth along the bottom, but this makes it a chore to take the cloths out for cleaning, and the 'over-and-over' lacing shown here, through the berth base, has much to commend it.

A leeboard for double berth

A cloth leeboard, which extends down the middle of a double berth, must be pulled tight down at the bottom. An eyeplate is needed for this beneath the mattress at each end. The thin rope that runs through the bottom seam of the leeboard is passed through each eyeplate, then back up to the tough eyelet at the end of the leeboard. After putting the line through this eyelet it is hauled tight and knotted. This keeps the leeboard well down and prevents it sliding up or down the berth.

Lashings at the top corners are secured to eyeplates on strong points about 4 feet (1.2 m) above the berth, and ideally there should be a central eyeplate above the middle of the berth to take the extra lashing.

This type of cloth leeboard is considered much more comfortable than a wooden one, especially as the latter tends to be so heavy and awkward to stow if it is high enough for all sea conditions.

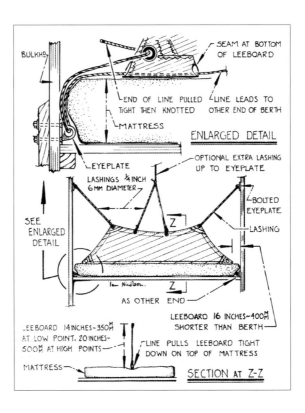

A wooden leeboard

The top sketch shows how to make a leeboard less shoulder-bruising by using a cushion. The cushion covering will normally match the mattress covering, but it will have flaps, like flanges, sewn along two edges. The cushion is held up and the top flap is screwed in place with 8-gauge countersunk screws at 6-inch (150-mm) intervals. Then the cushion is dropped in place and the bottom flap screwed with the same size screws at double the above spacing.

The sketch at the bottom left shows a leeboard with a hand-grip slot to make it easier to get into the berth. This slot, or row of slots, provides ventilation for the person in the berth if there is a high cloth leeboard. The sketch at the bottom right shows the traditional 'bottle-top' section, beloved of good boat builders because it looks superb when well made.

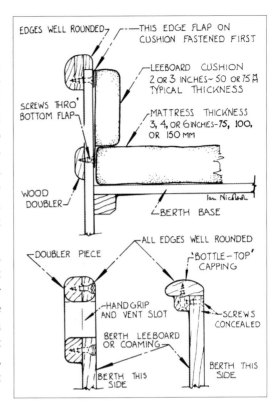

A cabin grab rail

For anyone who cannot do woodwork up to the standard of the furniture in his cabin, this grab rail is excellent. It also has the merit that it can be located anywhere under the cabin top deck, and two or even three of these ropes can be rigged. They can be taken down when the boat is in harbour, if, for instance, they obstruct the headroom, and they can be fitted over the galley, up forward, in a toilet compartment and so on.

The eyeplates, never eyebolts, at each end must not be the decorative type, which have little strength, but the type sold for securing on deck to take blocks and so on. To get the rope tight, a dinghy-type tackle, ideally with a built-in cleat or jammer, is used, or if preferred, a rigging screw can be fitted.

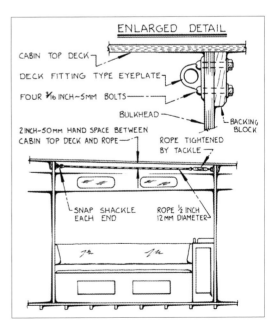

Handrails inside cabin tops

The upper sketch shows a shapely hardwood handrail, the lower one a metal rail, which can be of steel or aluminium alloy, or for someone who wants a boat to be different, polished brass could be used ... at a high cost. The cabin coamings are shown as being of fibreglass, but they can be of any material. If they are of fibreglass, the fastenings look best if sunk just slightly, then covered over with a carefully matched skin of gelcoat.

These designs have concealed fastenings, which will always enhance the appearance of any boat. The wooden longitudinal part must be epoxy-glued to the chocks carefully, and the screws must be at least 12-gauge to get a good grip. These screws should penetrate ⅝ inch (15 mm) into the longitudinal rail for the maximum grasp onto that component.

These designs can be used in other locations on board. It is good sense to standardise on items like handrails so that a whole batch for a boat can be made at one time, even though the handrails throughout the boat will be of different lengths.

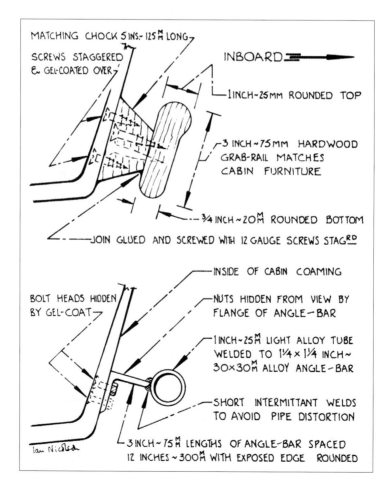

There are never enough handrails on board

A shortage of handrails below decks and on deck is the prime cause of bruises and even broken limbs. Handrails on the cabin top backed by similar rails on the underside make a lot of sense. If screws are used to hold the rails, as shown in the top sketch, the rails need to be slightly staggered. This means that the screws in the top rails do not foul those driven upwards from below, yet each set of screws is in the middle of its solid section of rail, between the handgrips.

Bolts are widely considered to be the best fastenings for handrails. They should not be of brass unless they are more than 5⁄16-inch (8-mm) diameter, and galvanized steel should be avoided as it tends to rust after a few years unless the end sealing is perfect. Nowadays, stainless steel is the common material.

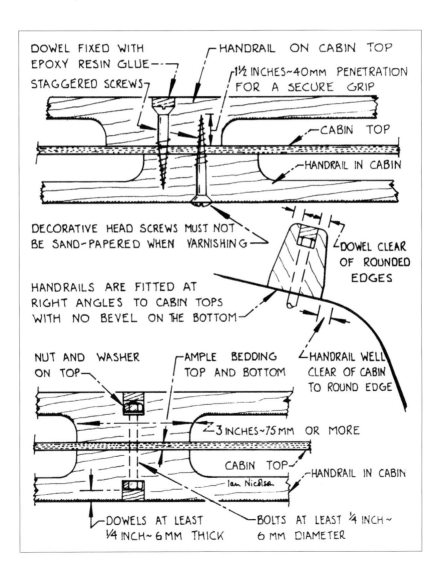

DOWEL FIXED WITH EPOXY RESIN GLUE

HANDRAIL ON CABIN TOP

STAGGERED SCREWS

1½ INCHES~40mm PENETRATION FOR A SECURE GRIP

CABIN TOP

HANDRAIL IN CABIN

DECORATIVE HEAD SCREWS MUST NOT BE SAND-PAPERED WHEN VARNISHING

DOWEL CLEAR OF ROUNDED EDGES

HANDRAILS ARE FITTED AT RIGHT ANGLES TO CABIN TOPS WITH NO BEVEL ON THE BOTTOM

NUT AND WASHER ON TOP

AMPLE BEDDING TOP AND BOTTOM

HANDRAIL WELL CLEAR OF CABIN TO ROUND EDGE

3 INCHES~75 mm OR MORE

CABIN TOP

HANDRAIL IN CABIN

Ian Nicolsa

DOWELS AT LEAST ¼ INCH~6 MM THICK

BOLTS AT LEAST ¼ INCH~ 6 MM DIAMETER

Better coat hangers

It is not possible to buy seagoing coat hangers, so they have to be made. They have to be 'captive', otherwise they come adrift from their top bar and deposit the clothes in a huddle at the bottom of the clothes locker. This type of hanger, made from ½-inch (12-mm) marine ply or ¾-inch (20-mm) solid wood, with well-rounded edges, is fitted onto the suspension bar before the latter is fitted into the clothes locker.

The top edges may be padded, and the hole for ties and belts is optional. Some owners will prefer just the single slot for trousers, but then, here as elsewhere, when making special fittings for a boat the professional or amateur shipwright can enjoy using his imagination as well as his skills.

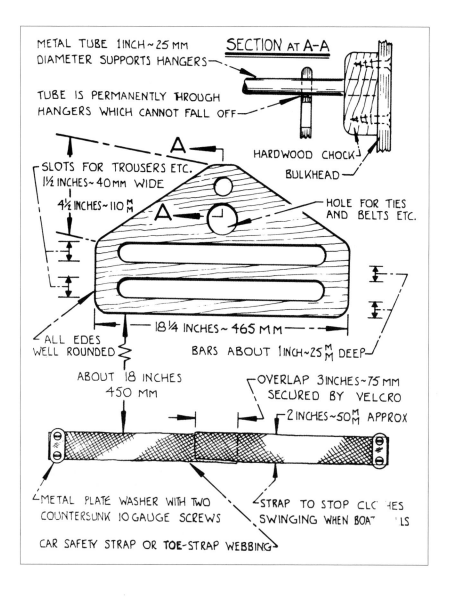

Seagoing drawers

An unbreakable handle on a drawer is all too rare. The one shown here consists of a slot through the front, which has a wood doubler over, partly to compensate for lost strength, partly to give a more pleasant grip, partly to improve the look of the drawer front. The size of the slot will accept the biggest, toughest seaman's hand, and it also allows ample air to flow in and out of the drawer to keep the contents dry and sweet-smelling.

To prevent the drawer flying open in a rough sea there is a turn-button, or maybe two, one at each top corner. These can be cut out of hardwood and may have some elegant pattern, perhaps fish or mermaids, carved on the face. It is no good screwing turn-buttons in place, nor are thin, flimsy bolts acceptable.

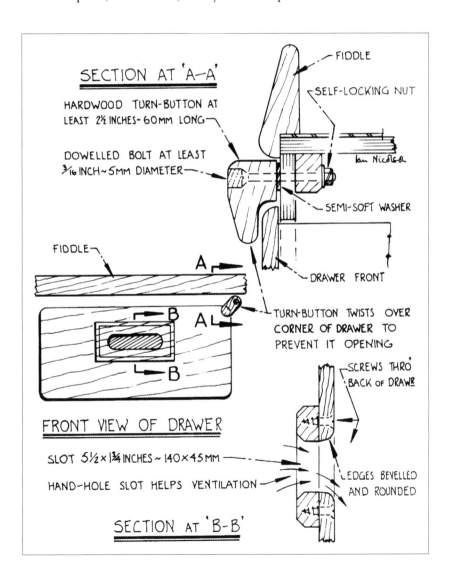

SECTION AT 'A–A'

HARDWOOD TURN-BUTTON AT LEAST 2½ INCHES~60MM LONG

DOWELLED BOLT AT LEAST ³⁄₁₆ INCH~5MM DIAMETER

FIDDLE

SELF-LOCKING NUT

Ian Nicolson

SEMI-SOFT WASHER

DRAWER FRONT

FIDDLE

A

B

A

B

TURN-BUTTON TWISTS OVER CORNER OF DRAWER TO PREVENT IT OPENING

SCREWS THRO' BACK OF DRAWR

FRONT VIEW OF DRAWER

SLOT 5½ × 1¾ INCHES ~ 140×45MM

HAND-HOLE SLOT HELPS VENTILATION

EDGES BEVELLED AND ROUNDED

SECTION AT 'B–B'

Safe drawers

It is common practice to make boat drawers with a lip on the bottom, to prevent accidental opening. On modern craft, which are usually light and lively, drawers too often jump up and burst open because the lip is inadequate for holding the drawer shut.

Two techniques are shown here which hold drawers shut with certainty. Top right is the well-tried swinging triangular metal plate. Anyone who wants to add a touch of traditional elegance to a boat can use polished brass, but most people will prefer stainless steel. Top left is the other reliable fitting: a standard barrel bolt that engages in the top flange of the drawer.

To stop the drawer coming right out when pulled hard, a string at the back has much to commend it, being cheap, adjustable and reliable. To prevent gear being thrown out of the back of the drawer when the boat is well heeled, there is a cover over the end quarter or one-third of the drawer length.

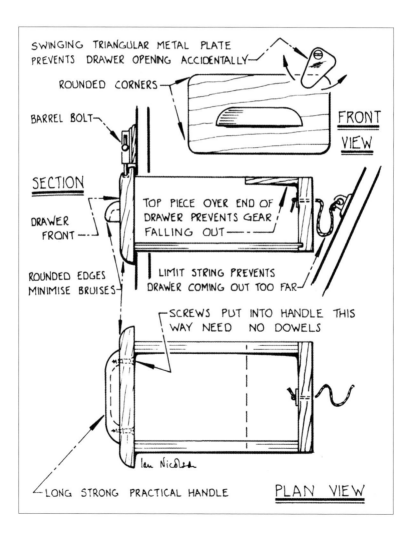

Bunk-side table

The pinnacle of luxury is lying in a snug berth sipping a mug of tea. If there is no bunk-side table, much of the pleasure is lost as the hot mug has to be balanced precariously. And if there is nowhere to put a plate of biscuits by the mug, then misery creeps in where there can be bliss.

A portable table that clips onto the berth's wooden leeboard has all sorts of advantages. It can be made quite large, to take a book or two and a transistor radio, but it will not obtrude in a cramped cabin because it can be stowed away when not in use. With luck, the same table can also be used elsewhere in the boat, perhaps on a partial bulkhead in the saloon or looped over a cockpit coaming.

Construction can be varied to suit available materials and skills. Anyone who is not good at making up wooden boxes can use a plastic box as the basis, and stiffen the side that takes the metal hooks with a doubler of wood or metal. Another panel is fixed to the bottom with small bolts to give extra stiffness and that extension which lodges against the leeboard side.

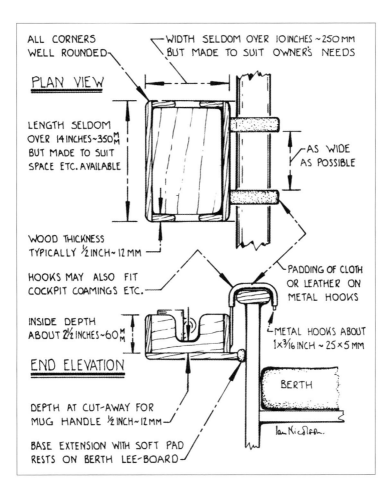

ALL CORNERS WELL ROUNDED

WIDTH SELDOM OVER 10 INCHES ~ 250 MM BUT MADE TO SUIT OWNER'S NEEDS

PLAN VIEW

LENGTH SELDOM OVER 14 INCHES ~ 350 MM BUT MADE TO SUIT SPACE ETC. AVAILABLE

AS WIDE AS POSSIBLE

WOOD THICKNESS TYPICALLY ½ INCH ~ 12 MM

HOOKS MAY ALSO FIT COCKPIT COAMINGS ETC.

PADDING OF CLOTH OR LEATHER ON METAL HOOKS

INSIDE DEPTH ABOUT 2½ INCHES ~ 60 MM

METAL HOOKS ABOUT 1 x 3/16 INCH ~ 25 x 5 MM

END ELEVATION

BERTH

DEPTH AT CUT-AWAY FOR MUG HANDLE ½ INCH ~ 12 MM

BASE EXTENSION WITH SOFT PAD RESTS ON BERTH LEE-BOARD

Stable table

Gimballed tables have lots of disadvantages: they are costly; they sometimes fail to keep pace with the boat's motion and build up a jerky roll, which chucks everything off the table; also they weigh a lot. Worst of all, when gimballing, no one can put a hand on them or they will pivot over and topple the dishes onto a settee.

This tilting table gets over a lot of the above disadvantages. Once the boat has been settled onto a tack the table is pivoted till it is nearly level by pulling the drop-nose pin out of the telescopic holder and putting it back in a new position with the table roughly horizontal. The table has to be designed so that it does not rise too high on the lee side, otherwise no one will be able to eat off it; on the other side, it must not close the knee space between the table and the settee.

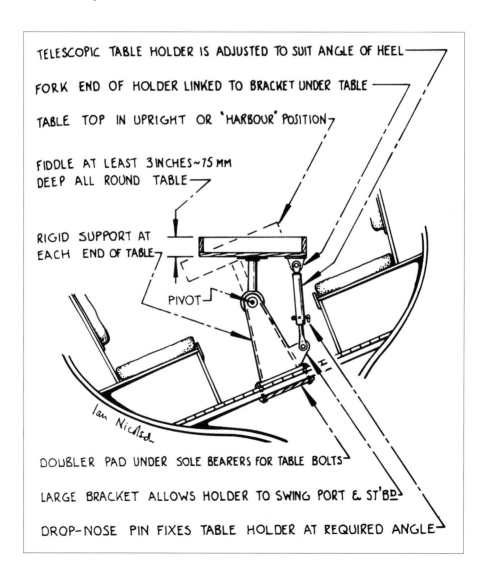

TELESCOPIC TABLE HOLDER IS ADJUSTED TO SUIT ANGLE OF HEEL

FORK END OF HOLDER LINKED TO BRACKET UNDER TABLE

TABLE TOP IN UPRIGHT OR 'HARBOUR' POSITION

FIDDLE AT LEAST 3 INCHES ~ 75 MM DEEP ALL ROUND TABLE

RIGID SUPPORT AT EACH END OF TABLE

PIVOT

Ian Nicolsd

DOUBLER PAD UNDER SOLE BEARERS FOR TABLE BOLTS

LARGE BRACKET ALLOWS HOLDER TO SWING PORT & ST'BD

DROP-NOSE PIN FIXES TABLE HOLDER AT REQUIRED ANGLE

A saloon table

If the main framework of a table is made in the form of a box, as shown here, the inside of the box can be used for stowing crockery, cutlery, bottles and so on. Naturally, the box must have a lid, and in this sketch the table top is pierced by two lids that do not extend out to the sides of the box, or to the ends. The lid on the left has deep fiddles secured to one side so that when, as shown at the top, the table is flush, it is suitable for writing or chart work. In the middle sketch the lid has been reversed so that the fiddles stand up, and they are shown cradling two bottles and two glasses.

The right-hand lid has a metal cutlery container fixed on the underside. The container need not be this shape, though this one has all sorts of advantages like strength, elegance and reliability, and it can be made of wood, metal or fibreglass.

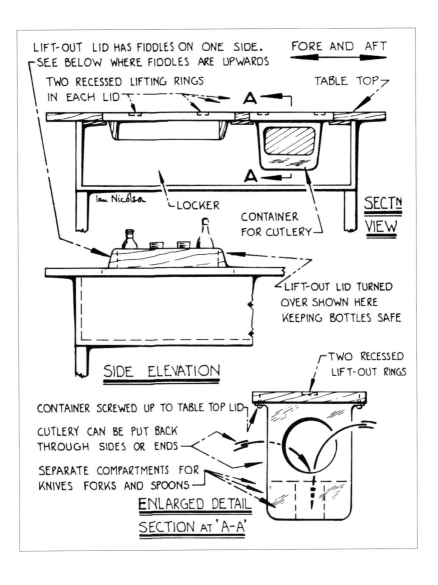

CHAPTER 5

Seats and Stools

A gimballed seat

For the single-hander or the ocean-cruising crew there is nothing so luxurious, so desirable or so comforting as a safe gimballed seat, which can be used in all but the very worst weather.

All sorts of designs can be used, and this drawing shows just some of the features that should be incorporated in a safe, strong and versatile seat.

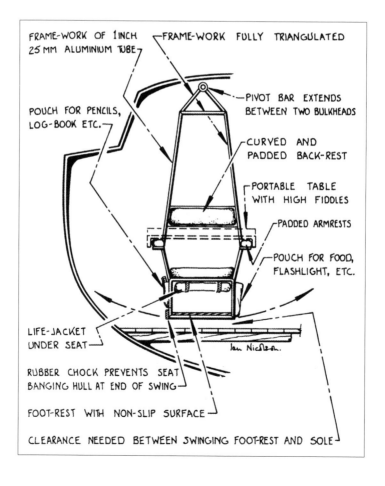

FRAME-WORK OF 1 INCH 25 MM ALUMINIUM TUBE

FRAME-WORK FULLY TRIANGULATED

POUCH FOR PENCILS, LOG-BOOK ETC.

PIVOT BAR EXTENDS BETWEEN TWO BULKHEADS

CURVED AND PADDED BACK-REST

PORTABLE TABLE WITH HIGH FIDDLES

PADDED ARMRESTS

POUCH FOR FOOD, FLASHLIGHT, ETC.

LIFE-JACKET UNDER SEAT

Ian Nicolson.

RUBBER CHOCK PREVENTS SEAT BANGING HULL AT END OF SWING

FOOT-REST WITH NON-SLIP SURFACE

CLEARANCE NEEDED BETWEEN SWINGING FOOT-REST AND SOLE

There must be ample swinging room, but some sort of end stop is needed port and starboard to deal with excessive heel. A lock is essential to keep the seat secure when gimballing is not required. Since this seat will be used for navigation, reading, eating and so on, a portable table is important. Without a footrest the seat will not be comfortable, but the rest should be close to the sole.

Almost an armchair

On the *Hedonist 44* there is a pair of adjacent chairs to this clever design. The most striking feature is the back, which is made up of curved tapered strips of hardwood carefully polished. The rest of the chair is of the same material so that quite apart from being a comfortable place to sit down, each seat is a striking piece of furniture.

This sort of seat hardly suits a boat under about 35 feet (11 m) long, unless she is very much a floating home, but for a deep-sea cruiser or a comfortable motorboat it has much to recommend it. The total space it takes up is not large, and it can be positioned by a desk or chart table provided that there is ample room to get in.

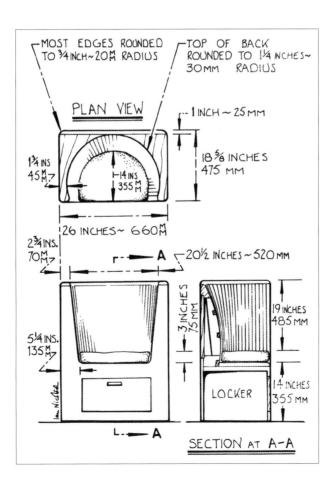

Perhaps the principal disadvantage of this form of seat is that it must either face fore-and-aft, or it may be uncomfortable on one tack because the sitter will be tipped out. On the opposite tack it will be bliss.

Comfortable seats

These 'offsets' show the shapes of two comfortable seats. They were designed as a result of research work done by V. Burandt and E. Grandjean and will fit most people. The top of the backrest should be curved to a radius of 650 mm, that is 25½ inches, and the middle of the backrest curved to a radius of 450 mm, which is 17½ inches.

Both seats have been modified to suit boat conditions, and the bottom one will be the easiest for most people to make as it does not require a contoured seat cushion, just a constant thickness foam rubber cover all over.

These seats can be made from fibreglass, cold-moulded wood, solid wood, or metal. The double curvature of the back is specially designed to give the correct support for the human backbone.

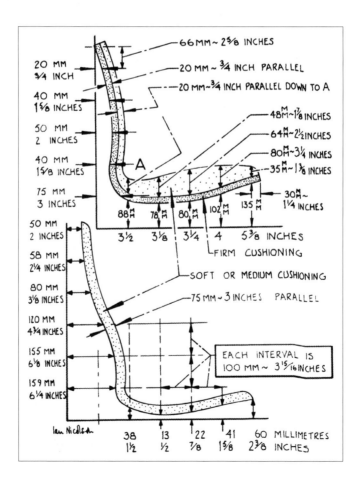

A stool for galley or chart table

A stool that swings to different positions, or swings out of the way when not in use, is ideal for many locations. The top can be made of solid wood without a cushion, or as shown here. The metalwork may be of steel for cheapness, aluminium alloy for lightness, or polished bronze for 'cost-is-no-worry' elegance.

There should be an average height with variations above and below for smaller and larger people. Most people like a seat diameter of about 10 inches (250 mm) but some boats are so cramped that an 8-inch (200-mm) diameter may be better. All wood and metal edges are well rounded, and the bracket bolted to the furniture must have six well-spread ¼-inch (6-mm) bolts, with a backing pad behind the support panel.

That split pin, shown on the enlarged detail, centre left, is to stop the seat lifting off accidentally at sea, and the plastic washers under each gudgeon ensure a smooth swinging motion when the seat is pivoted.

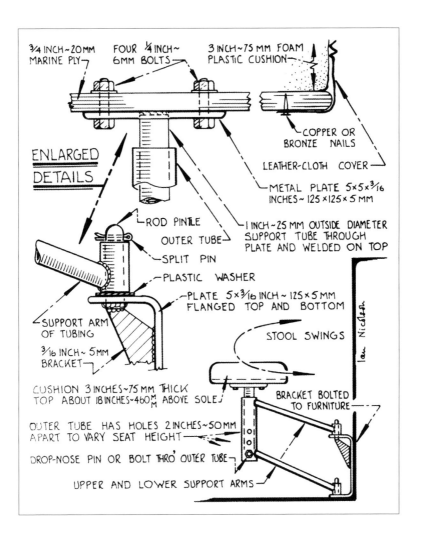

A portable stool

The cook and the navigator need a seat some of the time, as does the owner's wife when she is making-up in front of her mirror – if the boat is big enough for a dressing table. This stool has the virtues of simplicity and steadiness. Its four legs are well splayed out and they each have a bronze peg that lodges firmly in a brass plate let into the cabin sole. There can be several sets of these plates so that the stool can be used in different positions.

By having the legs and ties of the same section, cutting and planing the wood is simplified. The seat frame can be the same, or of the same thickness but a greater depth than shown here.

The seat can be flat, but the cold-moulded curvaceous one shown is more comfortable and elegant. It can have a fitted cushion, in which case the edge framing on the ply is not needed because the ply is entirely covered by the cushion material.

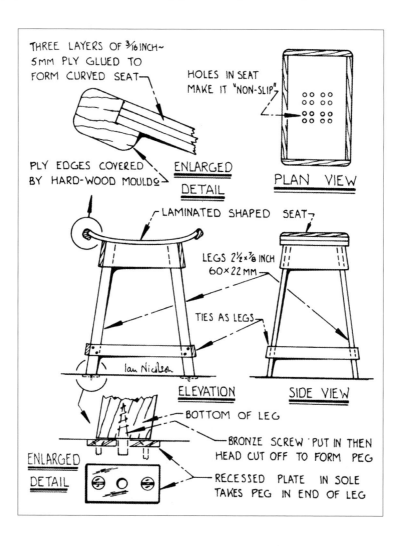

A seat for galley or chart table ... or both

It is a great advantage to be able to sit down to work, and this seat has all sorts of special assets. It stows away quickly and unobtrusively, like a drawer. When needed, its holding clip (not shown in the sketch) is released and it is simply pulled out. The end is pushed through a slot in the galley front and dropped down so that the lip prevents the seat sliding back.

Being narrow, it can be sat on in the ordinary way, or astride. It must be as long as the gap between galley and chart table, and if it is to have enough room to stow away horizontally it will be important to have the chart table fairly near the centreline of the yacht. The seat slides on runners like a drawer when it is pushed into its stowed position, and the end lip must be made so that it is easy to grip, or, alternatively, an end handle must be screwed on.

The section shows a light, strong assembly, but solid wood can be used, or a solid plank with a single stiffener or pair of them on the bottom. Whatever method is used to build up this seat, glue and metal fastenings are needed, and when ply is used the edges should be covered as shown.

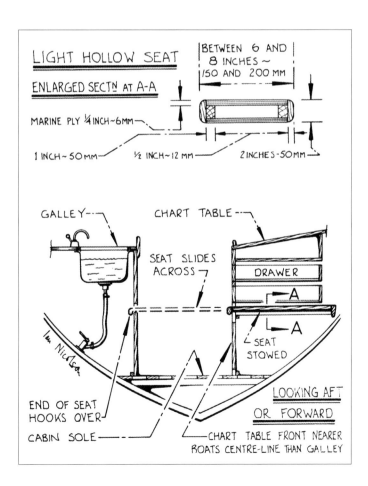

Curved seat for the navigator

Boats heel but navigators like to be upright, so a shaped seat by the chart table is popular. To keep the navigator safe, a bar that drops down and prevents him falling off the seat is a sensible precaution, provided it is strong enough to stand up to severe conditions. One end of this bar has a pivot bolt through a pair of short lengths of angle-bar on a bulkhead. The other end drops into a deep socket, or has a metal ring that hooks into a snap shackle or similar securing device.

Access to the locker under the seat should be from the side, otherwise the navigator has to be disturbed if the locker entrance is at the top.

To help the navigator stay in place, there is a toe-rail (like a wood fiddle) inboard of his footrest. This wood batten must stop short at the ends, otherwise a puddle may accumulate outboard of it; the end gaps are also helpful when cleaning.

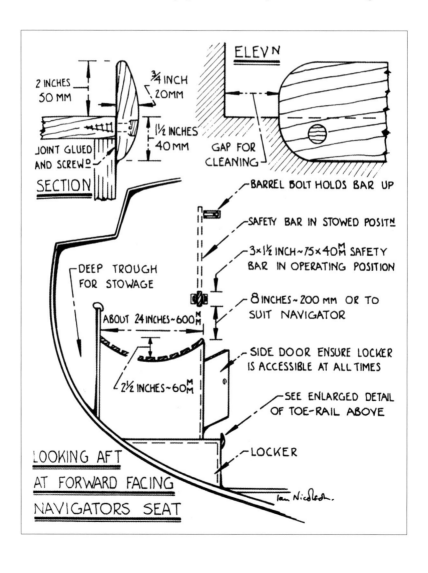

CHAPTER 6

The Toilet Compartment

Top yacht's toilet

The ideas here were seen in a Camper & Nicholson 46-footer. The WC is an electric one, hidden when not in use under a cork covered seat hinged down from the bulkhead. To deal with an electrical failure, the WC has a diaphragm-type pump concealed in a locker so that it is easy to get at when it needs servicing.

The adjacent locker has a shelf with an upper shelf to form a safe storage for cosmetics; this upper shelf is of Perspex and the cut-outs fit the shape of the different containers and bottles. If, after a few years or even months, different cosmetics are bought, a new Perspex panel with the appropriately shaped holes can be substituted.

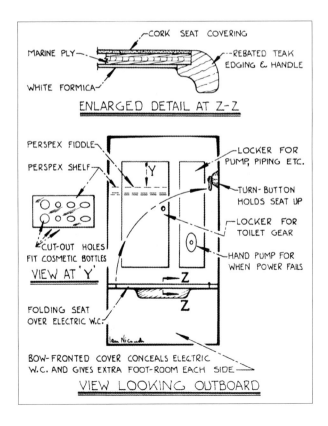

CORK SEAT COVERING

MARINE PLY

REBATED TEAK EDGING & HANDLE

WHITE FORMICA

ENLARGED DETAIL AT Z-Z

PERSPEX FIDDLE

PERSPEX SHELF

LOCKER FOR PUMP, PIPING ETC.

TURN-BUTTON HOLDS SEAT UP

LOCKER FOR TOILET GEAR

CUT-OUT HOLES FIT COSMETIC BOTTLES

HAND PUMP FOR WHEN POWER FAILS

VIEW AT 'Y'

FOLDING SEAT OVER ELECTRIC W.C.

BOW-FRONTED COVER CONCEALS ELECTRIC W.C. AND GIVES EXTRA FOOT-ROOM EACH SIDE

VIEW LOOKING OUTBOARD

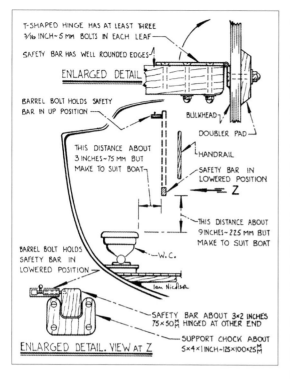

T-SHAPED HINGE HAS AT LEAST THREE 3/16 INCH~5MM BOLTS IN EACH LEAF

SAFETY BAR HAS WELL ROUNDED EDGES

ENLARGED DETAIL

BARREL BOLT HOLDS SAFETY BAR IN UP POSITION

BULKHEAD

DOUBLER PAD

THIS DISTANCE ABOUT 3 INCHES~75 MM BUT MAKE TO SUIT BOAT

HANDRAIL

SAFETY BAR IN LOWERED POSITION

Z

THIS DISTANCE ABOUT 9 INCHES~225 MM BUT MAKE TO SUIT BOAT

BARREL BOLT HOLDS SAFETY BAR IN LOWERED POSITION

W.C.

Ian Nicholson

SAFETY BAR ABOUT 3×2 INCHES 75×50 MM HINGED AT OTHER END

SUPPORT CHOCK ABOUT 5×4×1 INCH~125×100×25 MM

ENLARGED DETAIL. VIEW AT Z

Toilet guard bar

There is nothing funny about being pitched off the WC as the boat lurches over a wave top. This folding guard bar prevents that sort of accident.

It consists of a strong bar, here made of wood (though a metal tube can be used), with one end hinged to the bulkhead on one side of the WC. When in use the bar folds across to the opposite bulkhead, where its end rests in a recessed wood chock. To lock it down, there is a common barrel bolt, and another of these holds the bar in the upright, open position.

The precise position of the bar relative to the toilet seat height needs discovering by trial and error, as the height of the seat from the cabin sole has to be taken into account, as does the total space inside the toilet compartment.

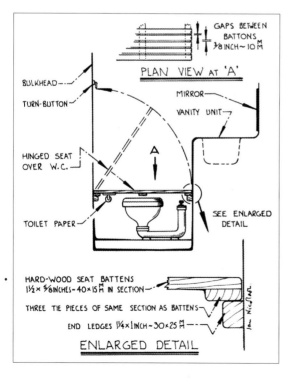

GAPS BETWEEN BATTONS 3/8 INCH~10 MM

PLAN VIEW AT 'A'

BULKHEAD

TURN-BUTTON

MIRROR

VANITY UNIT

HINGED SEAT OVER W.C.

A

SEE ENLARGED DETAIL

TOILET PAPER

HARD-WOOD SEAT BATTENS 1½×⅝ INCHES~40×15 MM IN SECTION

THREE TIE PIECES OF SAME SECTION AS BATTENS

END LEDGES 1¼×1 INCH~30×25 MM

ENLARGED DETAIL

Toilet compartment into boudoir

A small change can make an ordinary toilet cabin into a place where men can shave in comfort, and women can spend time making-up. The essential addition is a comfortable seat, and this may be fitted so that it hinges down over the WC. When the seat is up against the wall the WC is back in use and the toilet paper is handy, fixed to the underside of the hinged seat.

The seat, when it is in use, should be close to the top of the WC otherwise it will be too high for comfort, especially for small people. The mirror should extend right down to the level of the vanity unit and be 3 feet (1 m) high.

Tucked-away toilet

On a small boat, it's absurd to use up a great chunk of the limited available cabin space for the toilet. This compartment is only in use for a short time each day, and as it needs headroom, it tends to take up valuable space.

By putting the WC on a pair of slides, it can be pushed under the cockpit out of the way when not in use. Smells go up out of the main hatch swiftly, and the seacocks can be handy, the inlet being combined perhaps with a cockpit drain.

That folding step makes an excellent seat, but it must have a clip to hold it in the up position. Its top surface will probably be covered with Treadmaster or a similar slip-proof surface. The two lines of sail track on the sole are tough enough to stand up to all sorts of rough use, but will need a little lubrication. On the WC's wooden base there must be a barrel bolt which locks the WC in the stowed or in-use position as required.

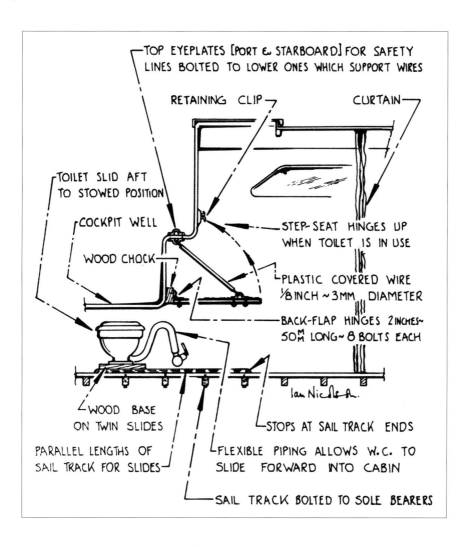

Pull-out basin over WC

One of the advantages of a basin arranged like a drawer is that it can be large, with ample space either side for washing gear, false teeth and so on. The drawer must have strong, smooth slides, and common sail track is ideal for this. To hold the basin, either when stowed away under the side deck or pulled out so that one of the crew can wash, there must be some locking device. Usually this is a simple barrel bolt, as shown top left.

Piping for the drain and water supply must be flexible and long enough to extend to the 'pulled-out' and 'closed' positions of the basin. It is usual to have the drain flattened at its bottom end and tucked under the WC seat, so when the basin plug is pulled out the dirty water runs into the WC bowl.

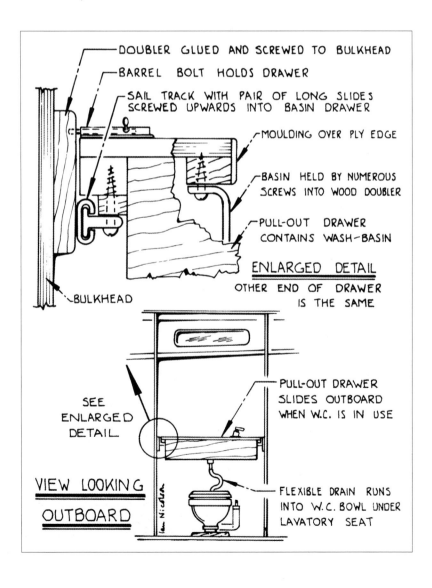

Keeping the toilet paper dry

The open-bottomed container for keeping toilet paper dry, even in gale conditions when drips come through the overhead vent, is stitched together by a sailmaker out of PVC cloth, so it never needs painting or varnishing. It weighs very little, and if anyone is flung against it in rough weather, they will not be bruised.

The horizontal rod can be made of metal tubing, or a wood or plastic bar. It can have a screwed-on end, as well as the fixed end, or the seamanlike 'split pin' shown here, made of a short length of thin rope with a decorative knot on top and a single half-hitch at the bottom to prevent the rope coming out by accident.

If the bulkhead is the usual thin ply, screws will either be too short, or will protrude right through. This explains why there has to be a backing block.

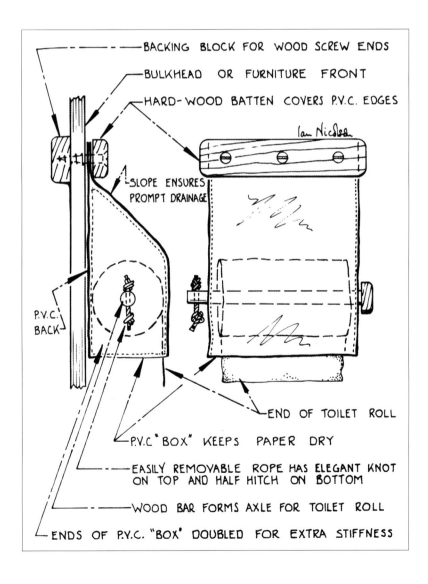

BACKING BLOCK FOR WOOD SCREW ENDS

BULKHEAD OR FURNITURE FRONT

HARD-WOOD BATTEN COVERS P.V.C. EDGES

Ian Nicolson

SLOPE ENSURES PROMPT DRAINAGE

P.V.C. BACK

END OF TOILET ROLL

P.V.C "BOX" KEEPS PAPER DRY

EASILY REMOVABLE ROPE HAS ELEGANT KNOT ON TOP AND HALF HITCH ON BOTTOM

WOOD BAR FORMS AXLE FOR TOILET ROLL

ENDS OF P.V.C. "BOX" DOUBLED FOR EXTRA STIFFNESS

CHAPTER 7

Plumbing and Cabin Heating

Emptying a wide flat bilge

In a boat with no deep sump in the bilge, water under the cabin sole slops about over a wide area. When the boat heels, the puddle of bilge water migrates outboard further and further as the angle of heel increases.

Two techniques are available to deal with this situation. The top drawing shows a single large bilge pump with suctions fixed athwartships at different points to deal with various angles of heel. The suctions are led to a manifold, which is a metal box with a row of valves, one for each suction. Normally, only one valve will be open at once.

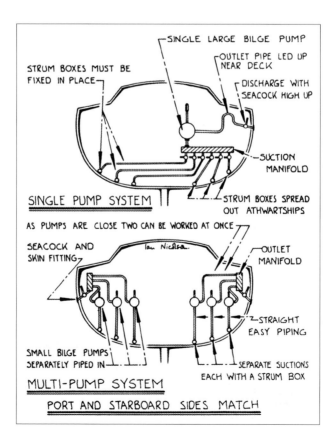

The second arrangement need not cost more, in fact, such is the cost of valves and longer runs of piping, it may be cheaper. A row of pumps is fixed on a bulkhead, port and starboard. Each pump has its own suction line and can have its own discharge, though in the drawing there are outlet manifolds each side, each taking three outlets. If there is a lot of water in the bilge, and if the person pumping does not have to hang on for safety with one hand, he can operate two of these small pumps at once.

Extra tankage

On any cruiser, it is almost impossible to have too much tank capacity. Often, when an old boat is examined it will be found that for every ten years of her life, the capacity of both fuel and water tanks has been increased by 20 per cent and sometimes even by 50 per cent. One way to get more capacity is to fit flat stainless steel tanks in otherwise unwanted spaces, such as below the cabin sole, at the bottom of the settee lockers, and in those areas too often made damp by bilge water.

These tanks should be firmly secured along the whole of two edges at least, with an absolute minimum of six bolts per tank. The bolts should be ¼-in (6-mm) diameter for every 20 feet (6 m) of boat length, and the tangs they go through should be at least twice as strong as the bolts. The tanks should also rest on semi-soft packing, such as wood or a dense foam plastic material – never on fibreglass.

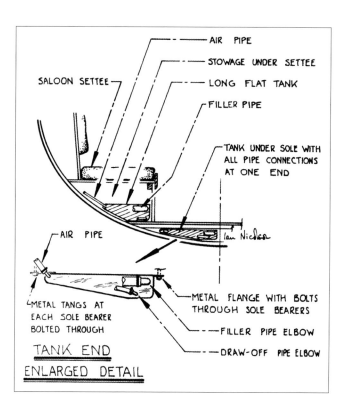

The important feature of this sort of tank is that for installation reasons, it is almost always best to have all the pipe connections at one end, and all in the form of elbows, so that the joining plastic pipes run off parallel with the end plate.

Fuel problems solved

When refuelling, it's hard to avoid some splashes round the filler pipe, and from there the contamination goes all over the boat, on deck and below. The preventative arrangement shown here is seen sometimes on large motor-cruisers but can be used on the smallest boat.

Each filler pipe is fitted in a stainless steel box with enough space around the pipe for old newspapers or rags to be spread. Drips during refuelling are blotted up by the paper or rags and disposed of ashore. The cap on the filler is fully watertight and is doubly protected as the surrounding box keeps water away from the filler. If water does get in, or if there is condensation in the drip well, the little drain hole located halfway up lets the water escape.

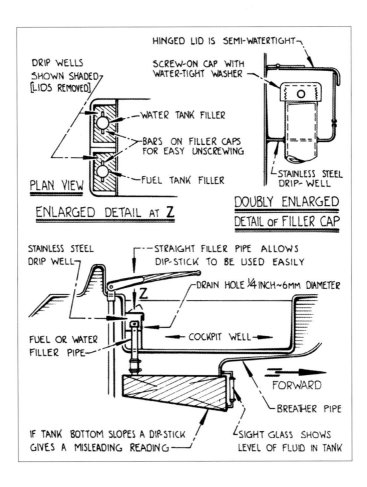

The same type of filler can be used for water, and if the pipe is kept straight it is easy to use a dipstick. Some people prefer a sight glass, as shown at the fore end of the tank, but of course both dipstick and sight glass have to be used intelligently if the fuel tank bottom slopes.

Clean refuelling

Fuel is smelly and it has a bad habit of staining whatever it touches. Wet fuel left on a deck attracts grit and it is hard to take fuel on board without spilling at least a few drips.

Having a special locker for fuel-filling pipes reduces a lot of complications, besides making it easier to keep a boat clean and pleasant. The locker may be let into the side of a cabin top, perhaps with an adjacent locker for the gas cylinders. When this locker is washed out, the water must drain overboard via a special pipe, which comes out below the waterline with its own seacock.

Washing out is not started until as much of the spilt fuel as possible has been mopped up with rags or paper. The filler pipe cap has to be put on before washing out starts, and even if a lot of water is swilling around inside the locker, none will get down this important pipe because it is raised about deck level. To make it easy to take the filler cap off, there is a stout rod welded across the top which is gently tapped to make the cap unscrew.

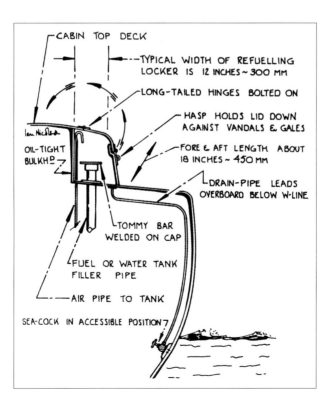

Solid fuel stove I

Nothing makes a cabin cosier than a good solid fuel stove. It can be kept burning low all day and all night, yet it gives out enough heat to warm a big area and may even feed a batch of radiators run from a back boiler.

All during the planning and fitting of a stove, the risk of fire has to be remembered. Also, a stove is heavy, so it needs a strong base that will stay put whichever way the yacht happens to tip. Before designing the base it is best to buy, or at least measure, the stove, so as to get the outside dimensions, the feet bolt centres, the location of the chimney relative to the front face and so on.

The base has to be secured to the hull with a big factor of safety (quite six or eight times) and the stove should be positioned so that a straight chimney can be installed, unless there is a very strong reason why this cannot be achieved. Ideally, the chimney should be in one length. No drying rail should be set low down, otherwise clothing may rest against the hot stove when the boat is heeling, and result in a fire.

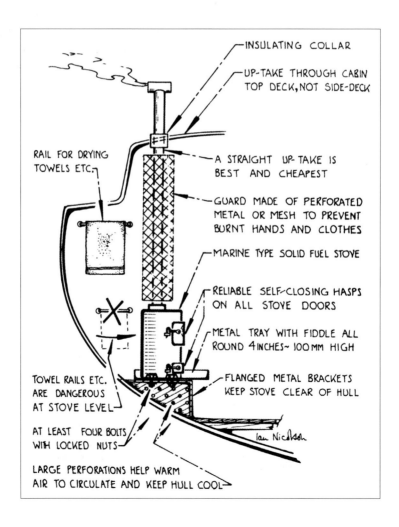

Solid fuel stove II

Insulating washers should be put under the stove feet to prevent heat reaching the support base. Nearby furniture and bulkheads must be protected either by distance, or heat-resistant materials. Another safety factor is the fire extinguisher, which should be just handy, but not right where a fire is likely to occur.

To use some of the ample heat that comes from a stove, a trunked fan can suck air from inside the chimney guard and feed it into the next cabin. This technique can be used on its own, or to supplement a back boiler feeding radiators in adjacent cabins.

The technique whereby the stove is in one cabin and the chimney led through another works well, but needs very careful insulation and is not cheap to set up.

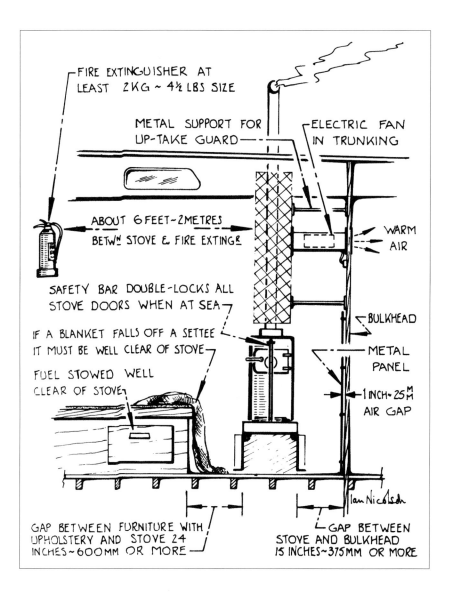

Chimney insulator

The chimney rising up from a solid fuel stove is hot, and can easily damage the cabin top deck unless there is some way of insulating the structure from the flue. An old-fashioned protector is the water trough, welded round the chimney. The trough is kept full of seawater and this evaporates, so it has to be replenished every so often. Replenishment may be by a drip feed from a tank on deck, or by a hose from a marina tap, set to drip into the trough.

In wet weather rain will fill the trough, but possibly not fast enough. At sea it may be a problem refilling the trough, unless there is enough spray slicing in over the weather rail to top up the water as it evaporates.

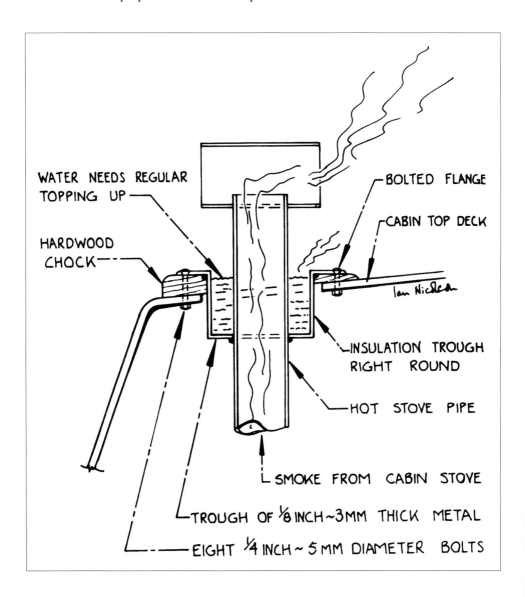

WATER NEEDS REGULAR
TOPPING UP

HARDWOOD
CHOCK

BOLTED FLANGE

CABIN TOP DECK

Ian Nicolson

INSULATION TROUGH
RIGHT ROUND

HOT STOVE PIPE

SMOKE FROM CABIN STOVE

TROUGH OF ⅛ INCH ~ 3MM THICK METAL

EIGHT ¼ INCH ~ 5MM DIAMETER BOLTS

Simple central heating

The attraction of this cabin heating arrangement is that it is easily made up using standard components that are widely available and cheap. To keep corrosion down, fresh water should be used in the circuit, and the components should not be of mild steel, except perhaps the holding tank. If this is of steel it should be galvanized or painted using epoxy resins.

Apart from short joining lengths, the piping should not be of plastic as this softens when hot water passes through it. Stainless steel, brass or copper piping can be used, and can be made up into radiators by bending in a zigzag manner.

The heating unit must be able to stand up to cooking on the galley stove. If the coil is kept flat it may be possible to cook a meal at the same time and on the same burner. Once the water in the circuit has been heated it can be pumped round for some time, gently warming the cabin, without further heating. The best flow rate, like every other aspect of this system, will need to be discovered by experimentation, and will depend on the size of the circuit, the pipe diameter, and so on. But in general, the smallest size of standard yacht pump is likely to be adequate.

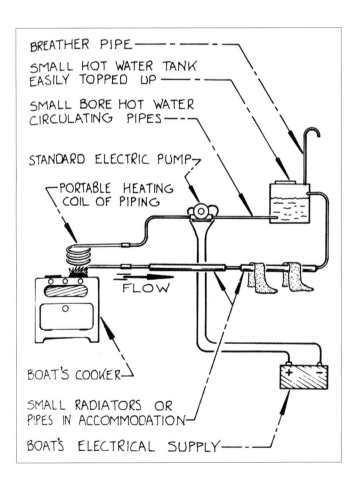

CHAPTER 8

Stowage

Stowage by the companion way

At the cabin exit everyone needs simple stowage racks for those items that are needed in a hurry. Different people have varying ideas about the equipment that should be stowed here, but the articles which are illustrated cover many crises. The fiddle with cut-outs is a shelf set above the main shelf, with sawn-out shapes to fit the tools etc.

The best kind of rack has room for additional things in future years. A deep shelf underneath for lengths of line, extra sail tiers or maybe a bar of chocolate and so on, is also a good idea. If the fuse box, with its numerous switches, is also built into the same unit, this will often save money and make this piece of furniture look purposeful, sensible and decorative.

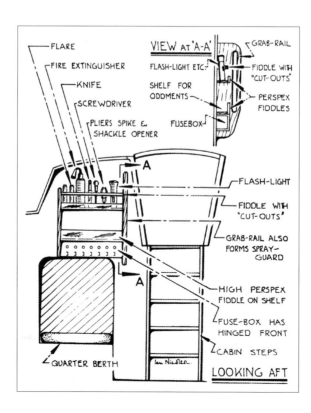

Making the inside edge of the shelves into a grab rail and protector against errant spray is common sense. The grab rail could be continued further down to help anyone wriggling into the quarter berth.

Stowage round a quarter berth

A shelf over the foot of a quarter berth uses up otherwise unwanted space. The shelf must not be so low that it fouls the feet of the sleeper, nor should it extend so far forward that it makes turning over on the berth impossible. A low fiddle is needed to keep kitbags on the shelf, as shown at top right. If the fiddle is just glued then no dowelling is needed. Vent holes are advisable on the shelf. The sides of the shelf are supported by glassed-over triangular section pieces, shown top left. The shelf is only screwed each side at the forward end, so it is easy to take out for cleaning. Forward of the shelf, there is space usually for a book rack or high fiddled shelf, and at the top of the berth a large bag secured to the side of the boat gives a lot of storage. If clothes are kept in this waterproof bag, the sleeper can lie comfortably against it even when the boat is banging about in a seaway.

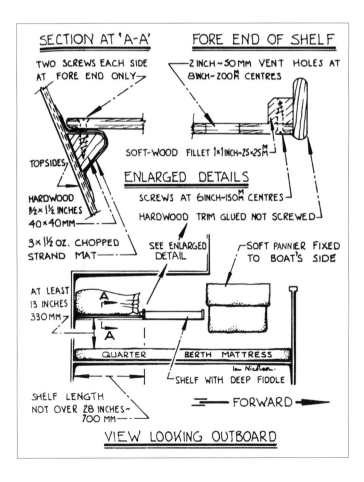

Big bottles

The trend towards buying drink in bigger bottles can make life difficult for anyone arranging suitable stowage space. However, the lower part of many clothes hanging lockers can be used without stealing important capacity because it is at the top of these lockers that a good fore-and-aft length is needed. A particular advantage of this bottle rack is that it is safe in rough weather, yet the hooch is close at hand and does not have to be delved for in deep, dark, dank caverns.

Normally, the bottle rack will be made off the boat and fitted with a wood rim all round, which is bolted or screwed to the bulkhead. The access hole must not be so large that it weakens a strength bulkhead and it should have rounded corners with a radius of at least 3 inches (75 mm). The slots, if large enough for 1½-litre bottles, will also hold big Coca Cola bottles, soda water containers and so on.

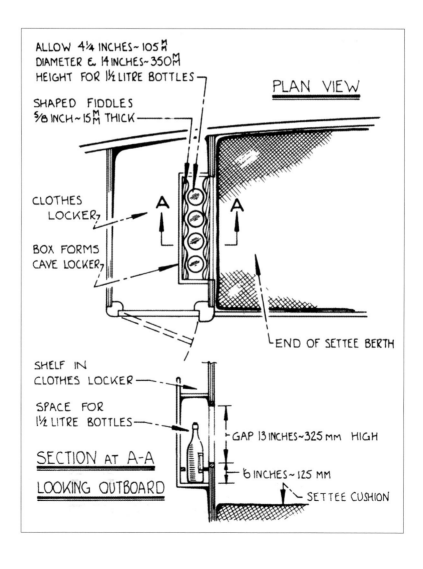

ALLOW 4¼ INCHES ~ 105 ᴹᴹ
DIAMETER & 14 INCHES ~ 350 ᴹᴹ
HEIGHT FOR 1½ LITRE BOTTLES

PLAN VIEW

SHAPED FIDDLES
⅝ INCH ~ 15 ᴹᴹ THICK

CLOTHES LOCKER

BOX FORMS CAVE LOCKER

END OF SETTEE BERTH

SHELF IN CLOTHES LOCKER

SPACE FOR 1½ LITRE BOTTLES

GAP 13 INCHES ~ 325 mm HIGH

6 INCHES ~ 125 MM

SECTION AT A-A LOOKING OUTBOARD

SETTEE CUSHION

Stowage between double bulkheads

It is usual for the bulkhead between the saloon and toilet to support the mast, so a double bulkhead here for extra strength is a special asset. It can also act as a sound barrier, which plenty of owners and their girlfriends favour, for extra privacy.

If the gap between the bulkheads is made into lockers and shelves, some should open forwards, some aft, to get the best use from the space. It makes sense for each member for the crew to have his own personal locker for toilet gear near the basin, and everyone wants a drip-proof locker for the toilet roll, but there is a limit to the amount of gear that can be conveniently stowed in the toilet compartment.

The top of the space can be used as a ventilator box, and part of it can be a vent trunk to the bilge or a chimney for the cabin stove.

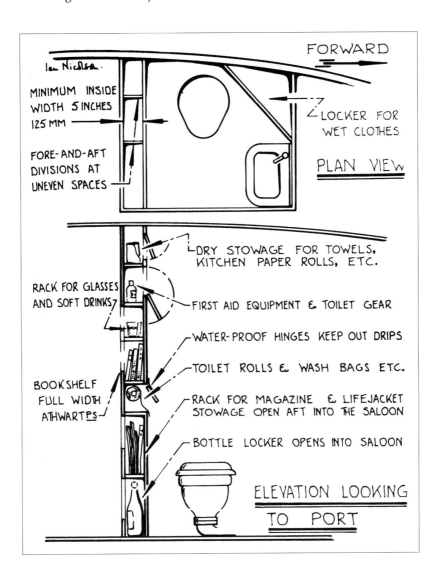

Unobtrusive stowage

It is common practice to fix fire extinguishers onto bulkheads, where they can be seen easily. Sometimes other safety gear, such as big flashlights, those outsized foghorns and even life jackets, are secured up in prominent positions where they can be grabbed in a hurry. If there is a shortage of bulkhead space, or if the owner wants to keep all his weight low down, or if gear screwed up on a bulkhead is going to bruise the crew in rough weather, alternative sites have to be found.

One good ploy is to indent the front of a settee or berth and have the safety gear secured in brackets in the recess. Here it is out of the way but easily reached. The space used is small, and the cost of arranging this type of stowage is minimal. Only if the equipment is heavy is it necessary to have support pillars behind the recess.

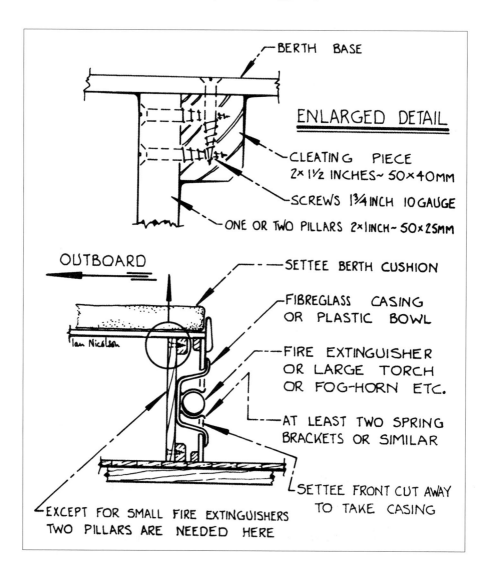

Dry under-berth stowage

Plastic boxes are cheap, easy to clean, available in many sizes and easy to replace. They make excellent storage containers because they are watertight and light. For use under berth tops, the kind with a lip all round is easy to fit using either rebated wood, or wood glued and bolted up in steps, as shown in the enlarged detail. It is usual to buy the biggest size of box that will fit the available space, and have a row of them the full length of the cavity under the berth. Bilge water can slop up in large dollops without wetting the gear stowed in the boxes. If the berth base is of fibreglass and a lot of it is cut away for access to the boxes, some compensating structure must be added to prevent the berth top collapsing when a heavy person sits on it. This reinforcing can consist of battens secured along the underside, or props as shown here.

Alternatively, a combination of props and stringers can be used.

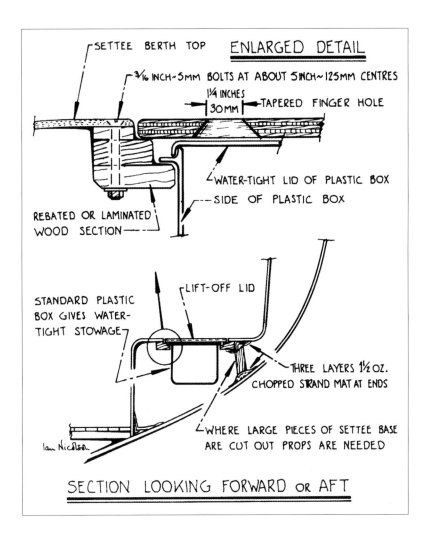

SETTEE BERTH TOP — ENLARGED DETAIL

3/16 INCH~5mm BOLTS AT ABOUT 5INCH~125mm CENTRES

1¼ INCHES 30mm — TAPERED FINGER HOLE

WATER-TIGHT LID OF PLASTIC BOX

SIDE OF PLASTIC BOX

REBATED OR LAMINATED WOOD SECTION

STANDARD PLASTIC BOX GIVES WATER-TIGHT STOWAGE

LIFT-OFF LID

THREE LAYERS 1½ OZ. CHOPPED STRAND MAT AT ENDS

WHERE LARGE PIECES OF SETTEE BASE ARE CUT OUT PROPS ARE NEEDED

Ian Nicolson

SECTION LOOKING FORWARD OR AFT

Stowage bins above the bilge water

This is yet another way of using watertight plastic boxes to keep gear dry. Even if the boat inverts, clothes, food, batteries and so on should stay safe in that strong polythene style of box, which has a tightly fitting lid.

By having pairs or sets of three or even four angle brackets under each box, the space beneath can still be used for gear which can get wet without worry. However, some shipwrights will prefer to make a fore-and-aft support, using perhaps a plank set on edge, the top taking the load of the plastic boxes and the bottom lying snug on the inside of the hull. Such a plank would be glassed over with four runs of 1½ oz chopped strand mat along each side, swept well up and down the hull.

The securing of the boxes must be such that the through bolts do not tear through the plastic, so the internal wood pad is essential and should be nearly as deep as the box, and almost as long fore-and-aft.

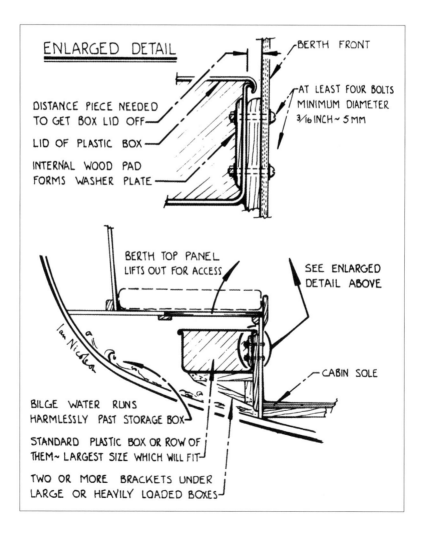

ENLARGED DETAIL

BERTH FRONT

DISTANCE PIECE NEEDED
TO GET BOX LID OFF

AT LEAST FOUR BOLTS
MINIMUM DIAMETER
3/16 INCH ~ 5 MM

LID OF PLASTIC BOX

INTERNAL WOOD PAD
FORMS WASHER PLATE

BERTH TOP PANEL
LIFTS OUT FOR ACCESS

SEE ENLARGED
DETAIL ABOVE

CABIN SOLE

BILGE WATER RUNS
HARMLESSLY PAST STORAGE BOX

STANDARD PLASTIC BOX OR ROW OF
THEM ~ LARGEST SIZE WHICH WILL FIT

TWO OR MORE BRACKETS UNDER
LARGE OR HEAVILY LOADED BOXES

Containers below settee

Lift off a settee cushion, take away the locker lids, and there is a large and very useful locker. Unfortunately, if there is even a tiny quantity of water in the bilge, it invades the settee locker when the boat heels, especially if the bilge is shallow.

To avoid getting wet gear, suspended containers are fitted under the settee top. These containers can be made of many materials, but PVC, as used to make sail covers, is a favourite. It is light, waterproof, easily made up by a local sailmaker, and it needs no painting or finishing treatment apart from the application of a waterproofing compound along the stitched joins.

In the top drawings, two alternative methods of securing the containers are shown. To use the one on the right, the container must have a slot each end for reaching through to hold the wood in place while the screws are put in. The left-hand sketch shows a technique whereby the wood strips are first bolted under the settee, then the container screwed to the wood, so no slots are needed in the container.

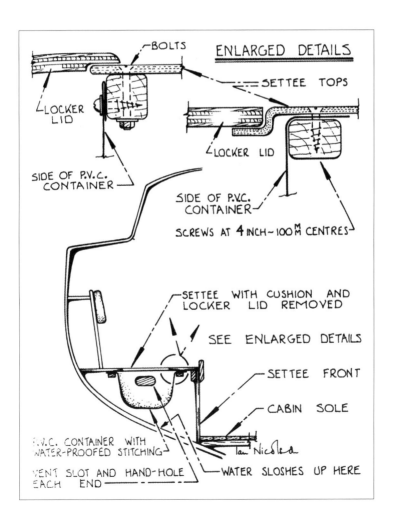

Under sole stowage

Stowage is seldom adequate in the galley, so though it smacks of desperation, there is justification for using the space below the sole as a store. Containers that fit exactly between the sole bearers usually have to be specially made, which explains why some owners have the bearers shifted to suit standard boxes and baskets.

The type of plastic-covered wire basket sold for deep freezes, shown top left, is popular though its ledges are at the ends, which usually means special lips may have to be fitted each side to lodge on flanges on the bearers. These flanges should be of hardwood, glued and screwed at 5-inch (125-mm) intervals.

Some people use plywood panels fixed on the undersides of the bearers, with round holes cut in the ply to take standard plastic buckets.

It always pays to buy the containers first, before making up the arrangements to support them. The hatches in the sole must be fairly small and easily portable. If there is one big sole hatch the cook will have great difficulty picking it up to get at the food stowed below.

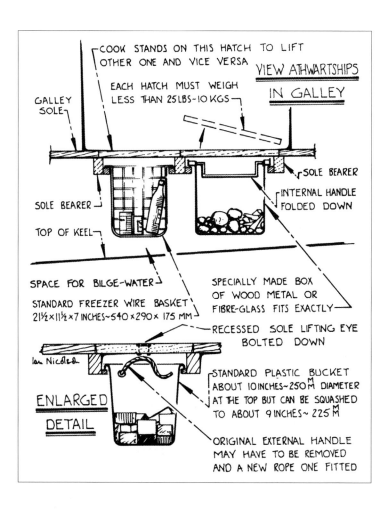

Quick and cheap extra stowage

There are few schemes that give cheaper extra stowage than the fitting of plastic boxes inside locker doors. These boxes can also be located under settees; in cockpit lockers beneath the seats; fixed inside oilskin lockers high up, where there is often a little space clear of the hanging oilskins, and so on.

Plastic boxes are available in a great variety of sizes from hardware stores, camping shops, industrial suppliers and so on. They need no painting and are waterproof.

Because the material is soft, there must by a wood batten to take the screws or bolts that support each box. A second wood batten along the bottom is only needed if there are to be lots of heavy items in the box, or if the box is longer than about 12 inches (300 mm). If in doubt, fit a base support to be safe.

Before fitting one or more of these boxes inside a locker door, the hinges should be checked. Usually, the weak point here is the size of the screws, and by changing to bolts the hinges can be made to support the extra load.

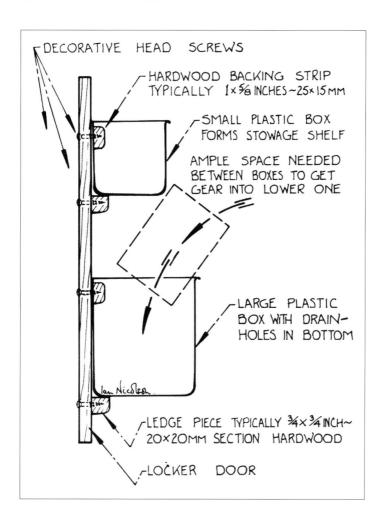

Fresh food storage

In the days before fridges, fresh vegetables and fruit were stored in special boxes on deck like this one. Sometimes there was a second box for fresh meat. The two essentials of this type of larder are ample ventilation and insulation. Thick wood was the secret of the insulation. Traditionally, these deck lockers were of varnished teak on expensive craft, and varnished mahogany on less costly boats.

They look delightful and can contain a lot of food, though it is important not to pack in too much otherwise air cannot circulate inside. The lids need clipping down to prevent them being blown open in a gale, and each box needs four short lengths of brass angle bar, one piece at each corner, with a bronze broad headed bolt to fix the box firmly to the deck.

This type of container relieves the congestion in the galley lockers. It can be used for stowing all sorts of food in waterproof containers but there must be protection from hot sunlight, by an awning or tent over the boom.

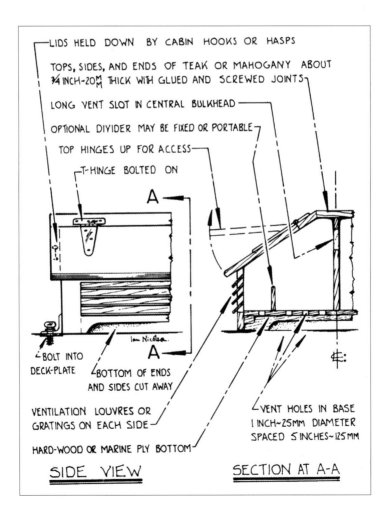

LIDS HELD DOWN BY CABIN HOOKS OR HASPS

TOPS, SIDES, AND ENDS OF TEAK OR MAHOGANY ABOUT ¾ INCH-20MM THICK WITH GLUED AND SCREWED JOINTS

LONG VENT SLOT IN CENTRAL BULKHEAD

OPTIONAL DIVIDER MAY BE FIXED OR PORTABLE

TOP HINGES UP FOR ACCESS

T-HINGE BOLTED ON

A

A

BOLT INTO DECK-PLATE

BOTTOM OF ENDS AND SIDES CUT AWAY

VENTILATION LOUVRES OR GRATINGS ON EACH SIDE

HARD-WOOD OR MARINE PLY BOTTOM

VENT HOLES IN BASE 1 INCH-25MM DIAMETER SPACED 5 INCHES-125MM

SIDE VIEW   SECTION AT A-A

Plastic pipe stowage containers

All over the world, it is easy to buy strong plastic piping from builders' merchants. This tubing makes excellent watertight containers when one end has been bunged up, and the other fitted with a watertight cover. These tubes can be used to convey gear safely to a cruiser even when it is raining hard; they keep gear dry even in wet lockers; they can be used to hold emergency equipment for use if the yacht sinks, and are easy to take into a life raft.

A wood-end bung is a cheap way of sealing off a tube, but some builders' merchants sell flanges, which can have ply pads bolted over. Occasionally, it is possible to buy pipe-end seals. The end where the gear goes in can be closed off by common plastic bags with ordinary elastic bands, but the arrangement drawn is better. If ordinary plastic bags are used, the first should be put on and sealed with two elastic bands, then the second and so on.

If these tubes are fixed under the thwarts of a dinghy, they give buoyancy as well as providing dry stowage. Aluminium tubes or fibreglass ones can be used just as well as plastic piping.

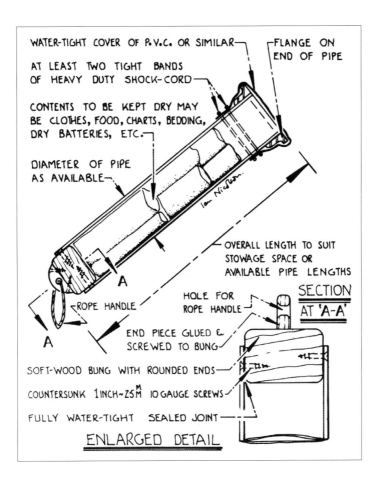

WATER-TIGHT COVER OF P.V.C. OR SIMILAR

FLANGE ON END OF PIPE

AT LEAST TWO TIGHT BANDS OF HEAVY DUTY SHOCK-CORD

CONTENTS TO BE KEPT DRY MAY BE CLOTHES, FOOD, CHARTS, BEDDING, DRY BATTERIES, ETC.

DIAMETER OF PIPE AS AVAILABLE

OVERALL LENGTH TO SUIT STOWAGE SPACE OR AVAILABLE PIPE LENGTHS

SECTION AT 'A-A'

HOLE FOR ROPE HANDLE

ROPE HANDLE

END PIECE GLUED & SCREWED TO BUNG

SOFT-WOOD BUNG WITH ROUNDED ENDS

COUNTERSUNK 1 INCH~25^M^M 10 GAUGE SCREWS

FULLY WATER-TIGHT SEALED JOINT

ENLARGED DETAIL

CHAPTER 9

Fiddles

Padded fiddles reduce bruises

Modern boats are light and therefore lively. They jump about and so they throw their crews about. To minimise injuries there must be no exposed sharp edges, and it helps if furniture and other projecting structure is padded.

The padding can be very simple, as shown here. Common foam plastic cushion, bought in strip form or cut from a chunk, is glued in place and covered with a suitable cloth. On tough, practical, no-nonsense craft the covering may be canvas, Terylene or similar material. Often it is the same material that is used to cover the cushions, so it may be a man-made suede, a leather cloth, or one of those tough practical new materials that stand up to so much hard work.

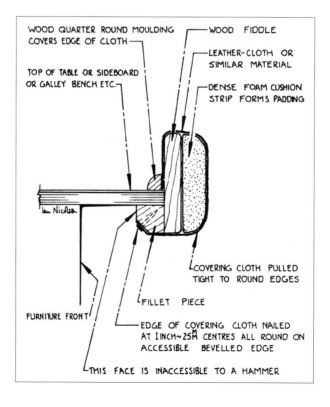

WOOD QUARTER ROUND MOULDING COVERS EDGE OF CLOTH

WOOD FIDDLE

LEATHER-CLOTH OR SIMILAR MATERIAL

TOP OF TABLE OR SIDEBOARD OR GALLEY BENCH ETC.

DENSE FOAM CUSHION STRIP FORMS PADDING

Ian Nicolson

COVERING CLOTH PULLED TIGHT TO ROUND EDGES

FILLET PIECE

FURNITURE FRONT

EDGE OF COVERING CLOTH NAILED AT 1INCH~25M CENTRES ALL ROUND ON ACCESSIBLE BEVELLED EDGE

THIS FACE IS INACCESSIBLE TO A HAMMER

The way the fillet piece under the table top is bevelled, so that the joiner can hammer in a row of nails, is noteworthy. These nails are totally concealed, and they must be non-ferrous otherwise they will soon rust.

Fold-away fiddle

In a quiet harbour the cook likes to work away at all sorts of interesting concoctions and high fiddles round the galley get in the way. At sea, the absence of a full set of sensible fiddles makes cooking impossible in all but the calmest weather. The great attraction of folding fiddles is that they do not get lost during a long spell in harbour, and they are quickly flipped up into position when needed.

The design shown here is held with those strong cabin hooks that do not clink and rattle because they have tapered hook ends, which slide into tapered holes in the eyepiece. It is no good securing these hooks with screws because at times the cook will grab a fiddle to prevent being thrown across the cabin, and then the loads on all parts of the fiddle will be serious. Because of these high loadings, there must be close-spaced hinges held with the longest screws that can be fitted.

Perhaps the only disadvantage of this style of fiddle is that when folded down it prevents locker doors being opened unless their tops are set well below worktop level.

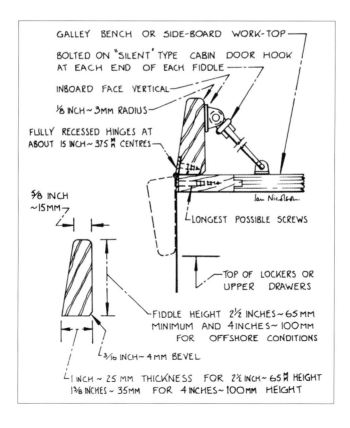

GALLEY BENCH OR SIDE-BOARD WORK-TOP

BOLTED ON "SILENT" TYPE CABIN DOOR HOOK AT EACH END OF EACH FIDDLE

INBOARD FACE VERTICAL

⅛ INCH ~ 3MM RADIUS

FULLY RECESSED HINGES AT ABOUT 15 INCH ~ 375 ᴹᴹ CENTRES

⅝ INCH ~15MM

LONGEST POSSIBLE SCREWS

Ian Nicolson.

TOP OF LOCKERS OR UPPER DRAWERS

FIDDLE HEIGHT 2½ INCHES ~ 65 MM MINIMUM AND 4 INCHES ~ 100MM FOR OFFSHORE CONDITIONS

3/16 INCH ~ 4 MM BEVEL

1 INCH ~ 25 MM THICKNESS FOR 2½ INCH ~ 65 ᴹᴹ HEIGHT 1⅜ INCHES ~ 35MM FOR 4 INCHES ~ 100MM HEIGHT

Fridge fiddle

If a fridge, or any similar cupboard-shaped container, faces athwartships, the contents will fall out when the door is opened on the wrong tack. What is needed is a fiddle on each shelf, but the insides of fridges are not designed to take fiddles. The metal or plastic interior will not easily accept fastenings, and sometimes it is impossible to join anything to the smooth interior lining.

A simple way round the problem is shown in this sketch. Each fiddle is made to jam tightly in place. On one end there is a rubber pad, and on the other a tightening device that presses firmly against the fridge side wall. Care is needed, otherwise the tightening nut may be turned so hard it will fracture the fridge lining. Once the bolt is pressing its rubber-tipped wooden nosepiece in place, the locking nut is tightened.

This fitting can be used in a dozen different locations, for instance, where there is fine polished furniture that must not be sullied by coarse carpentry.

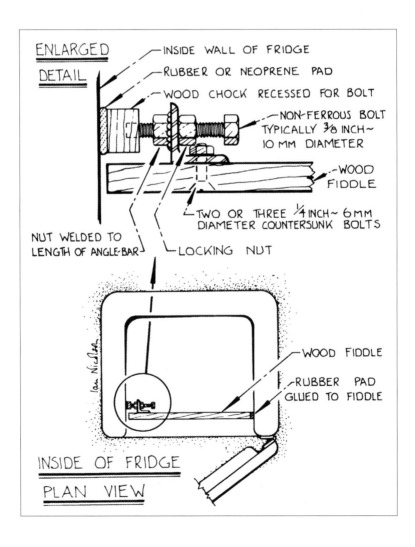

Fiddle for offshore cruising

Clear of the protection of land, boats jump about a lot even in moderate weather, and in severe conditions they cavort in every way, though hopefully not to the extent of standing on their head. Working surfaces like galley tops and sideboards need really high fiddles, but no boat (certainly no production boat) has these. This drawing shows a rugged and practical yet easily made fiddle, which will be a real asset in deep-sea conditions. It will double as a handgrip, provided it is toughly bolted in place, and it is easily and quickly made.

When the time comes to clean the galley thoroughly it will only take a few seconds to remove this fiddle, and in harbour it will be stowed out of the way, beneath a berth or in some similar out-of-the-way location.

The height and thickness of the fiddle will not change much regardless of the size of boat, because the former is dependent on the size of crockery, and the latter on the weight of a man hanging onto it.

The fiddle should be made of a hardwood and the finger grip should be glued and screwed to the main part, using 1¼-inch (30-mm) 8-gauge screws at 4-inch (100-mm) intervals.

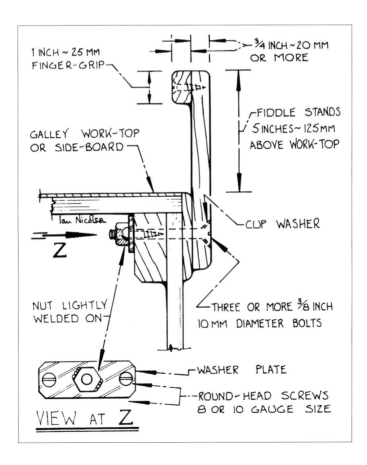

A semi-portable fiddle

Fiddles get in the way when the galley is being cleaned or painted, and they are not needed when the boat is in a sheltered berth. It is therefore logical to make them removable, though they must still be strong. The left side of this drawing shows the plan and elevation of a fiddle end fixed to a bulkhead by a pair of vertical slides, so that the fiddle drops into place just like a weatherboard. The end of the fiddle that drops in between the slides must be rectangular to ensure a snug fit.

At intervals along the fiddle there are bolts that can be put in without tools because the heads are knurled, and each nut has a little lever welded to it for hand tightening. The knurled top should have its upper edge well rounded, and it should be only about ¼ inch (6 mm) thick, so as to be unobtrusive.

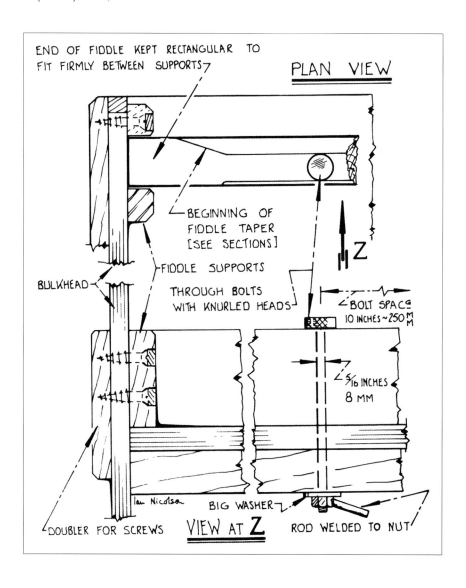

A semi-permanent fiddle

Hand slots in fiddles make life easier for the crew, but they must be high and short enough to prevent cutlery and the like from slipping through. The bottom right sketch shows how the fiddle is secured through the cleating piece, which joins the galley (or sideboard) top to its front. Butterfly nuts are shown, but these are not always easy to obtain, which explains why an alternative design is shown on the facing page.

Untapered fiddles look clumsy and unprofessional, but the bevelled side must be away from the worktop.

The whole of the fiddle is not tapered, a little 'flat' being left at the bottom to make it easy to grip the wood in a vice when working it.

In a boat which is embellished, the fiddle section shown bottom left may be preferred, even if it is slightly more difficult to clean.

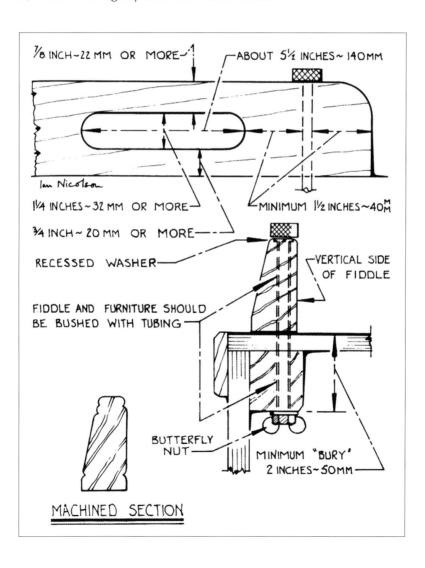

7/8 INCH ~ 22 MM OR MORE

ABOUT 5½ INCHES ~ 140MM

Ian Nicolson

1¼ INCHES ~ 32 MM OR MORE

¾ INCH ~ 20 MM OR MORE

MINIMUM 1½ INCHES ~ 40 M M

RECESSED WASHER

VERTICAL SIDE OF FIDDLE

FIDDLE AND FURNITURE SHOULD BE BUSHED WITH TUBING

BUTTERFLY NUT

MINIMUM "BURY" 2 INCHES ~ 50MM

MACHINED SECTION

CHAPTER 10

Bunks, Mattresses and Soft Furnishings

Offshore berth

This design of berth is comfortable when the boat is offshore in rough weather. It has the added advantages of being light, cheap, easily repaired and it occupies little space athwartships.

As the base is made of cloth, only a thin mattress is needed, and some people have been known to sleep well with no mattress. If a mattress is used, it is about one and a half times or even twice the width of the berth base. The height of the inboard leeboard

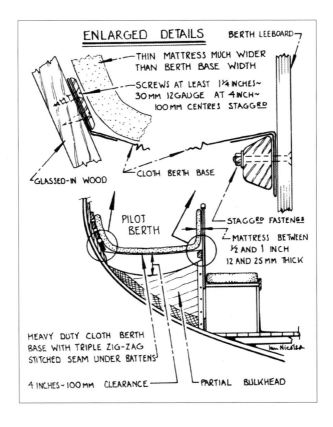

ENLARGED DETAILS   BERTH LEEBOARD

THIN MATTRESS MUCH WIDER THAN BERTH BASE WIDTH

SCREWS AT LEAST 1¼ INCHES~ 30 MM 12GAUGE AT 4 INCH~ 100 MM CENTRES STAGGRED

GLASSED-IN WOOD   CLOTH BERTH BASE

PILOT BERTH   STAGGRED FASTENGS

MATTRESS BETWEEN ½ AND 1 INCH 12 AND 25 MM THICK

HEAVY DUTY CLOTH BERTH BASE WITH TRIPLE ZIG-ZAG STITCHED SEAM UNDER BATTENS

4 INCHES~ 100 MM CLEARANCE   PARTIAL BULKHEAD

will probably be 15 inches (375 mm) above the berth base, and the battening on the outboard side will extend up roughly the same distance. A sailmaker will make up the base, using something like 10 oz Terylene (Dacron) sailcloth and a sewing machine that produces zigzag stitching. The berth width will usually be of the order of 21 inches (530 mm) wide for the shoulders and hips, and perhaps 13 inches (330 mm) at the foot.

Dual purpose pillow

The great majority of people like to have a pillow under their heads when sleeping, but ordinary 'shore' pillows are not ideal on a cruiser. The one shown here is made as part of the berth mattress, so it cannot slide off the berth. It is covered with the same waterproof material that keeps the mattress dry and easily cleaned.

Instead of filling the pillow with the usual type of wadding, this one has a thin top layer of the softest foam plastic cushioning material. Under it are spare sweaters, shirts and other clothes, which are individually wrapped in plastic bags so that they stay dry however much the boat takes in water through the hatches, window leaks, through the ventilators, and so on.

For extra comfort, the mattress has an upper layer of the extra soft foam plastic sponge material.

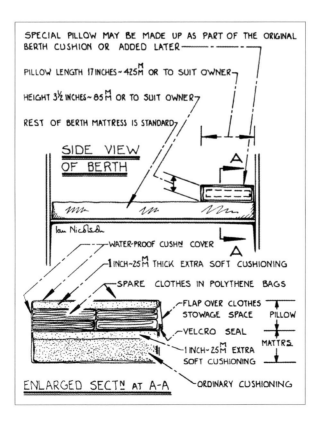

Engine access and quietness

Between a quarter berth and the engine bay it is usual to have a flimsy bulkhead, which makes access to the engine hard but does little to mute the rumble and shriek of the machinery. Instead of a ply panel, a curtain of PVC or similar waterproof material has a lot to recommend it. The curtain is rolled up and secured with three or four tapes while work goes on, servicing or repairing the engine.

The curtain is in two thicknesses, with some flexible sound-proofing material between. The noise-reducing material is often of foam plastic with a lead insert, so it must be in strips to allow the curtain to be rolled. Alternatively, the curtain can be a semi-rigid one with a hinge along a line above the soundproofing.

The stowage pouches are an optional extra, likely to be appreciated by the person who sleeps in the quarter berth as they provide safe, dry stowage in all weathers.

The top of the curtain is held by four or more bolts about 4-mm diameter through a hardwood batten in a seam along the top of the material. It is important that the curtain extends far enough to seal gaps between it and the ends of the engine space, and tapes or other securing arrangements will probably be needed here, spaced about every 6 or 10 inches (150 or 250 mm).

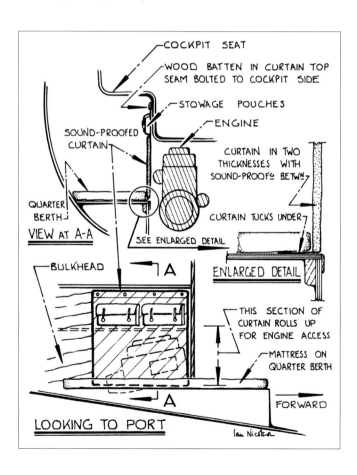

Staying dry in the quarter berth

It is hard to stop every drop of water coming in through the main hatch, even with a hood, and even when the hatch is kept shut most of the time. But a wet berth is such a catastrophe that it is worth going to considerable lengths to avoid it.

A good protective curtain is secured tightly along the top and it extends aft of the hatch, in this case under the bridge deck. The forward end wraps round the partial bulkhead and is held from top to bottom with Velcro. The bottom edge does not need securing, but it must reach well below the mattress.

With this sort of watertight curtain in place, anyone sleeping in the quarter berth is likely to get a headache eventually from lack of oxygen, which explains the carefully worked in airflow along the bunk.

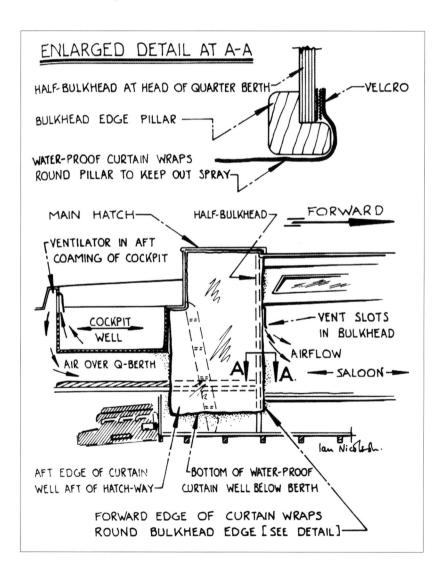

ENLARGED DETAIL AT A-A

HALF-BULKHEAD AT HEAD OF QUARTER BERTH — VELCRO

BULKHEAD EDGE PILLAR

WATER-PROOF CURTAIN WRAPS ROUND PILLAR TO KEEP OUT SPRAY

MAIN HATCH   HALF-BULKHEAD   FORWARD

VENTILATOR IN AFT COAMING OF COCKPIT

COCKPIT WELL

VENT SLOTS IN BULKHEAD

AIR OVER Q-BERTH

AIRFLOW

A   A.   SALOON

Ian Nicolson.

AFT EDGE OF CURTAIN WELL AFT OF HATCH-WAY

BOTTOM OF WATER-PROOF CURTAIN WELL BELOW BERTH

FORWARD EDGE OF CURTAIN WRAPS ROUND BULKHEAD EDGE [SEE DETAIL]

Avoiding bruises

The inside of a boat is full of sharp edges, which can bruise even when the boat is in harbour. If padding is fitted round those dangerous projections, the safety and comfort of the craft is enhanced enormously.

The top sketch shows a typical wood, fibreglass or rectangular section metal beam or bulkhead stiffener, or perhaps a mast support pillar. The padding should be as thick as practical, and it pays to make up a short experimental section to try in place before fabricating longer lengths. ½-inch (12 mm) thick foam is about the minimum thickness used, but it is hard to get anything as bulky as 1 inch (25 mm) thick to bend round sharp corners.

The middle sketch is a typical angle bar, such as is found in aluminium and steel craft. The bolts through the small flange must be put in first, then the padding is wrapped round and pulled up tight before the final row of bolts is fitted. It is important to have washers under the bolt heads.

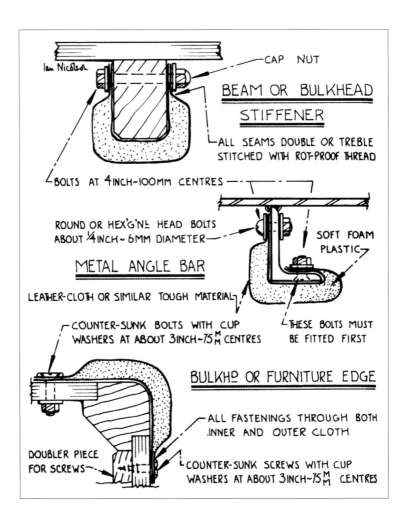

At the bottom is a typical furniture or bulkhead edge. Before fitting the padding, the sharp wooden edge should be planed away as much as possible. Where possible, bolts should be used, but if screws have to be put in they must be quite 1 inch long, so backing pieces of wood may be needed.

Padded doors and bulkheads

It is only in expensively finished boats that padded bulkheads are seen, yet the idea has merit for all sorts of craft. If anyone gets thrown about in rough conditions, it is much better to bump against a semi-soft surface. This type of bulkhead doubling also cuts down noise and helps to insulate cabins. The leather cloth finish will need regular wiping down in order to keep it clean, but this is more economical than re-varnishing or repainting every few years.

This sketch shows how the leather cloth edging is finished, and how the cloth is never taken round sharp edges. The solid chock in way of the door furniture is important, and a sensible detail is the 'breaking' or planing off of the edges of the solid chock.

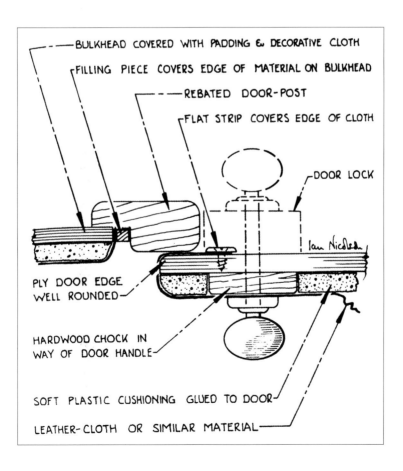

BULKHEAD COVERED WITH PADDING & DECORATIVE CLOTH

FILLING PIECE COVERS EDGE OF MATERIAL ON BULKHEAD

REBATED DOOR-POST

FLAT STRIP COVERS EDGE OF CLOTH

DOOR LOCK

Ian Nicolson

PLY DOOR EDGE WELL ROUNDED

HARDWOOD CHOCK IN WAY OF DOOR HANDLE

SOFT PLASTIC CUSHIONING GLUED TO DOOR

LEATHER-CLOTH OR SIMILAR MATERIAL

A padded pillar

Those mast support pillars that stand up in the middle of cabins can be bruising in rough weather. Whether the pillar is of wood or metal, it may look ugly and too much like a piece of scaffolding. A smart way to dress it up is to fit it with a jacket made from the same material as the upholstery, or cushion covers, or an attractive leather cloth ... or even real leather. If the jacket is padded, as shown in the section view, top right, it will be less bruising and have a pleasant feel, yet will still form a useful handrail.

It will be best to carry the jacket right to the deck head and down to the cabin sole, but not below it. The jacket can be held in place by a long Velcro strip concealed inside a fold-over flat, or by a full-length zip, also concealed, or a row of metal fasteners such as are used in dressmaking. But, of course, the hooks or press studs must be the non-rusting, heavy duty type.

A pleasant little extra is the row of long narrow pockets for pencils, near the top. After all, pencils are forever getting lost.

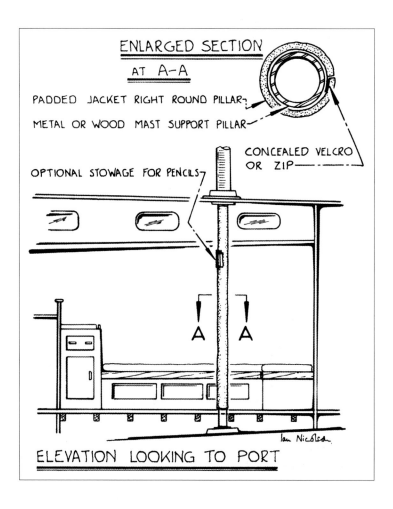

A curtain to avoid pinched fingers

A good reason for having a curtain instead of a door is that it cannot pinch fingers or cause other injuries. It gives less privacy, but provided the curtain is thick, perhaps double thickness, and provided it has a good overlap each side, it can be effective.

One way of adding to its effectiveness is to lace one side to a batten, which is either drilled with holes all along for the lashing, or scalloped out with a series of cuts to make it easy to pass the lashing through. Yet another technique is to screw one vertical edge of the curtain to the bulkhead, using roundhead screws with large washers under the heads, or even a thin wood batten.

Of course, this is a bit of a nuisance each autumn when the curtain has to be taken away for cleaning.

Using a length of sail track and slides along the top makes sense, because ordinary curtain rails are so weak and inclined to rust or corrode.

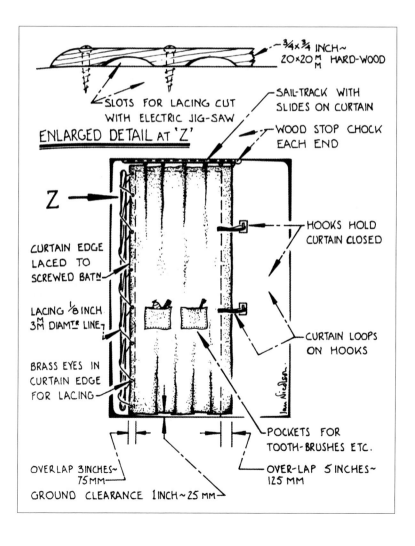

¾ × ¾ INCH ~
20×20 MM HARD-WOOD

SLOTS FOR LACING CUT
WITH ELECTRIC JIG-SAW

SAIL-TRACK WITH
SLIDES ON CURTAIN

ENLARGED DETAIL AT 'Z'

WOOD STOP CHOCK
EACH END

Z

HOOKS HOLD
CURTAIN CLOSED

CURTAIN EDGE
LACED TO
SCREWED BATN

LACING ⅛ INCH
3 MM DIAMTR LINE

CURTAIN LOOPS
ON HOOKS

BRASS EYES IN
CURTAIN EDGE
FOR LACING

POCKETS FOR
TOOTH-BRUSHES ETC.

OVERLAP 3 INCHES ~
75 MM

OVER-LAP 5 INCHES ~
125 MM

GROUND CLEARANCE 1 INCH ~ 25 MM

CHAPTER 11

Door and Drawer Clips and Catches

A truly tough door bolt

Items like hinges, locks and so on are called door 'furniture' in the marine world. They tend to be too weak to stand up to the rigours of life afloat, especially when the sea conditions are very bad, which is, of course, just when total reliability of every component on board is essential.

The owner who wants his locker and door bolts to be beyond reproach can make up this very simple fitting, using an attractive hardwood which matches the rest of the surrounding furniture. If the bolt is fitted inside a toilet compartment, the split

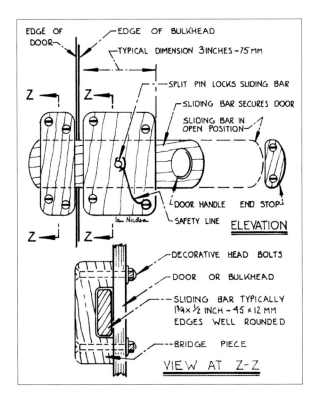

EDGE OF DOOR

EDGE OF BULKHEAD

TYPICAL DIMENSION 3 INCHES ~ 75 mm

SPLIT PIN LOCKS SLIDING BAR

SLIDING BAR SECURES DOOR

SLIDING BAR IN OPEN POSITION

DOOR HANDLE   END STOP

SAFETY LINE   ELEVATION

Ian Nicolson

DECORATIVE HEAD BOLTS

DOOR OR BULKHEAD

SLIDING BAR TYPICALLY 1¾ x ½ INCH ~ 45 x 12 MM EDGES WELL ROUNDED

BRIDGE PIECE

VIEW AT Z-Z

pin will be needed to lock the sliding bolt. There will be two holes in the sliding bar, one in the position shown, and the other located so that the sliding bar is secured in the open position.

Screws can be used through the door and bulkhead into the thicker bridge pieces, instead of the bolts shown.

A primitive but reliable door lock

It is a sad fact of life afloat that so few things go on working indefinitely. Even the common door bolt breaks, or the screws get loose, or the bolt is too flimsy to withstand a sudden lurch against the door, or corrosion gives trouble.

The door latch shown here can be made so big and tough that it never fails, and if it needs repairs or modifications, the job can be done easily with only a few common hand tools. The latch consists of a wooden bar pivoted at one end, with the other end dropping into a socket. There is a wooden knob on the latch to make life easy, and tied to this knob is a light line, which goes up and through a hole in the door to a loop on the

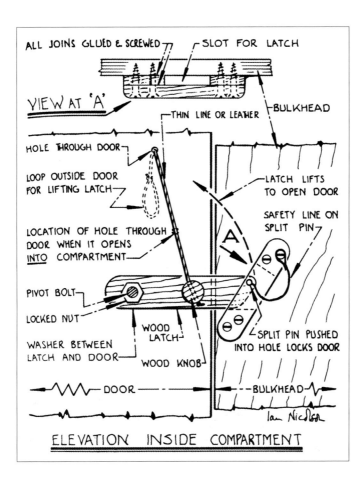

other end. To open the door from the outside, simply pull the loop down. To lock the door on the inside, just pop the split pin in the end of the latch to prevent it being lifted.

This simple gadget can be used on doors opening inwards or outwards. In the latter case, the latch end must be raised high enough to clear the bulkhead edge. Devices like this give great peace of mind because they are reliable.

A big, reliable turn-button

So many fittings, especially door and locker components, are too small, too flimsy and not able to stand up to the rigours of hard weather offshore. The turn-button shown here can be made easily and it should be indestructible, besides being easy to use with wet, cold hands.

On the locker door there is a wooden ramp or wedge, so that as the turn-button is twisted over it, the pressure on the door increases. The small sketch at the bottom right shows how the locker door is prevented from being pushed in too far by a rubber doorstop. This stop can be packed up simply by adding extra washers, and in this way wear and tear, warping and distortion are all taken care of. This door is always held tight and never rattles.

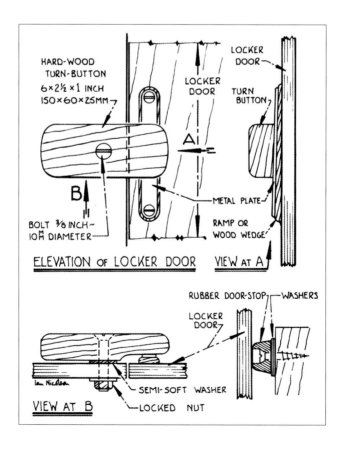

'Silent' type hook

For all sorts of jobs like holding ladders up or down, and doors closed, there is nothing to beat the simple cabin hook. But it must be the 'silent' type, because otherwise the fit of the hook into the eye is slightly sloppy, and the grip of the hook less than perfect.

A 'silent' hook never rattles, and unless the boat turns upside down it will not release its hold. It is ideal for such jobs as holding an engine casing door shut, because the same hook can engage in a second eye, to hold the door open when there is work to do on the engine.

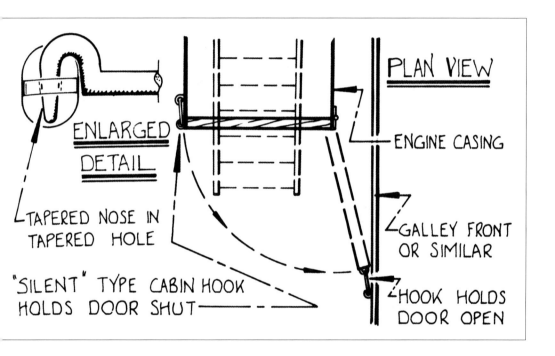

PLAN VIEW

ENLARGED DETAIL

ENGINE CASING

TAPERED NOSE IN TAPERED HOLE

GALLEY FRONT OR SIMILAR

"SILENT" TYPE CABIN HOOK HOLDS DOOR SHUT

HOOK HOLDS DOOR OPEN

Catches and clips

The top sketches show a type of metal-retaining catch that is used to hold up covers and lids which have horizontal hinge lines. These include chart table tops, which hinge up to allow charts to be taken out of the locker below, icebox and freezer lids, folding seats and so on. To prevent the lid banging back and forward between the bulkhead and the rotating plate when the boat rolls or pitches, the lid handle has a rubber knob on top and the handle almost fills the space for the lid. The rubber knob can be a doorstop.

At the bottom of the drawing is a simple pendulum plate, which pivots on a bolt. Where it is absolutely impossible to fit a bolt, a long, strong screw is used. The plate is normally bronze, brass or stainless steel, though thick Perspex may be used. It is a help to the crew if the contents of the drawer or locker are marked on the plate, even if the words have to be shortened. When the boat is in harbour, the plate is rotated anti-clockwise until it lies on top of the stop screw, and this lets the drawer or locker open unhindered.

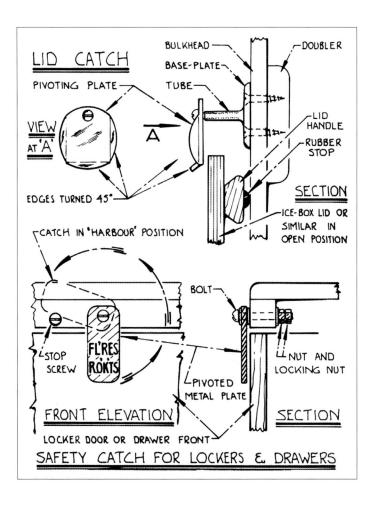

A lid clip

There are so many places on boats where a catch or retaining clip is needed to hold up a hinged cover, lid, or folding top. Typical examples are iceboxes that have top access, those chart tables which lift, hinge-up engine casings and cabin tables that hinge up to stow against a saloon bulkhead.

This neat, pivoted catch suits all these and many more situations. The one shown here is so simple that it can be made by an amateur with few tools to hand. A professional would probably machine a slot in the top flange of the angle bar and have the pivoted catch in the slot to give a symmetrical appearance. He would use a length of rod tapered each end with a ground-away weld as a pivot. The gadget shown here can be made to any size, to suit any weight of door or lid, and it should be proof against the worst conditions.

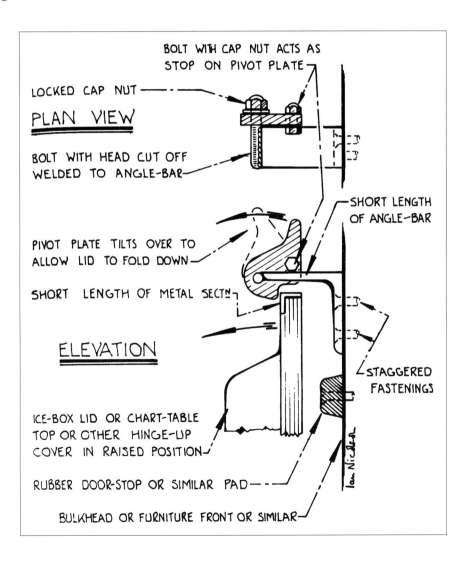

BOLT WITH CAP NUT ACTS AS STOP ON PIVOT PLATE

LOCKED CAP NUT

PLAN VIEW

BOLT WITH HEAD CUT OFF WELDED TO ANGLE-BAR

SHORT LENGTH OF ANGLE-BAR

PIVOT PLATE TILTS OVER TO ALLOW LID TO FOLD DOWN

SHORT LENGTH OF METAL SECTN

ELEVATION

STAGGERED FASTENINGS

ICE-BOX LID OR CHART-TABLE TOP OR OTHER HINGE-UP COVER IN RAISED POSITION

RUBBER DOOR-STOP OR SIMILAR PAD

BULKHEAD OR FURNITURE FRONT OR SIMILAR

Locker door stays

That thin brass chain that is used to pull the traditional type of lavatory cistern is ideal for preventing locker doors opening too far. A retaining chain is needed each side unless the locker door is under 10 inches (250 mm) wide.

To prevent the chains catching between the door and the locker front when the door is shut, the chains are angled in, and at the top they have chocks which ensure the whole chain is well in from the door sides.

Where possible, the chain end should be held by a bolt. Where this cannot be arranged, the end screw should be long, as thick as possible, and have a washer under the head to prevent the chain coming off.

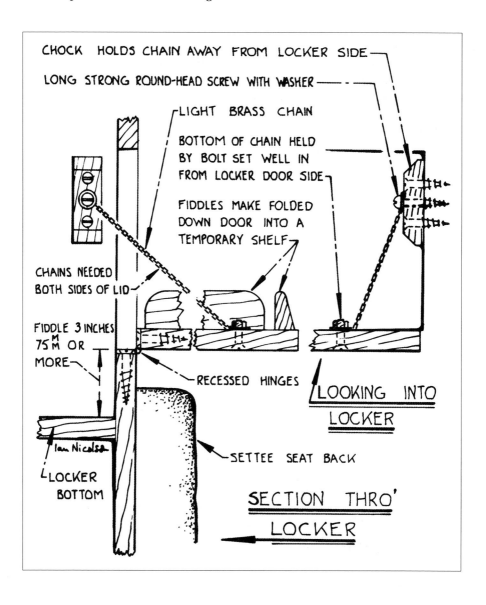

CHAPTER 12

Steps of Every Sort

Easily moved cabin steps

Nothing is as awkward as heavy cabin steps that have to be shifted to get at the engine, or lockers, or maybe the main fuse board. If the steps are portable, they cannot be pushed negligently forward into the cabin or they will damage the furniture and crew as the boat plunges about.

The arrangement shown here has the advantage that no one in the cockpit will try to come down the steps when they have been moved. Also, it is a fairly effortless job to lift the bottom of even heavy steps. The rope used to support the steps may be lashed to a hook or eye, which also supports the leeboard of a berth.

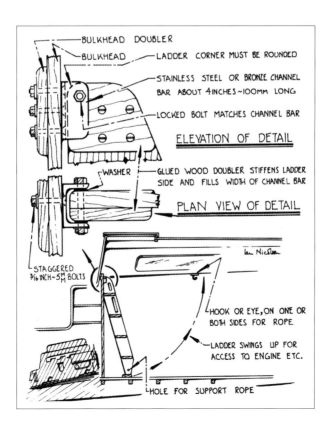

At the bottom of the steps there must be recessed wood chocks to keep the sides exactly in place. The top is hinged by pivot bolts through short lengths of channel bar, and this metal section should be bought before making the steps. Though full length steps are shown here, this technique can be used for shorter steps extending from an engine casing down to the cabin sole.

Sloped steps for sailing craft

When a yacht is heeled, the cabin steps are tilted and dangerous. By having the steps angled up each side, life is made easier for the crew. For added strength, the angled part of each tread may be continued down under the horizontal part, as shown here.

The top of the engine casing normally has no bevelled top area, but this is illogical, and the arrangement shown on the right makes a lot of sense. On the left, the usual type of fiddle is shown, though this important safety feature is often omitted or made flimsy and incapable of standing up to a heavy foot crashing down on it. It should be quite ¾ inch (20 mm) thick.

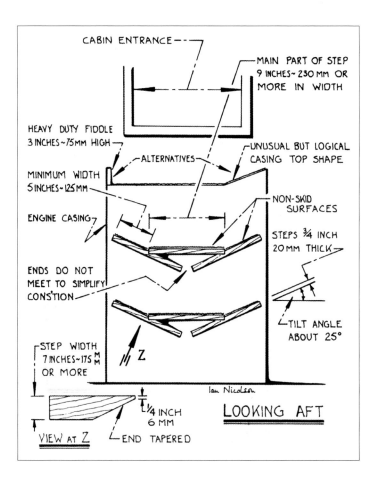

Though these steps are shown on the front of an engine casing, they can be fixed to a bulkhead or some other structure. They must be strongly secured with glue and at least three 12-gauge screws for each of the three parts of the step. An advantage of this design is that the side and centre portions form a triangle, strengthening each other.

Handrail and safety rail

It is common practice to have cabin steps that have to be taken away in order to gain access to the boat's engine. When steps have been removed, someone may not notice and step into the cabin with painful results. To prevent this type of accident, a handrail that can be swung across the cabin entrance is easy to make.

At the aft end of the handrail there is a pivot, and at the fore-end a clip that engages in a metal eye. The clip and eye can be set of spinnaker boom inner end fittings, and the same arrangement may be used at the aft end of the handrail. Indeed, the whole fitting may double as a spinnaker bearing out spar, used to prevent the spinnaker sheet cutting into the windward shrouds and getting so far inboard that the sheet winch cannot haul the boom aft.

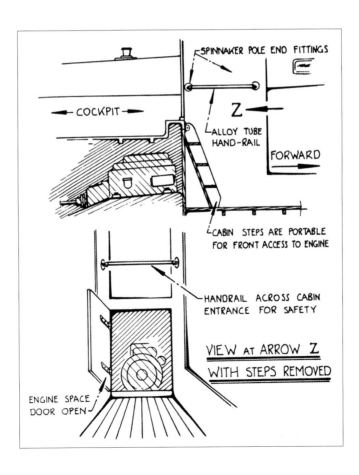

Though the handrail in the top sketch is shown to be horizontal, it can run down parallel with the steps. If there is an aft cabin, the same type of swing-across handrail can be used there too.

Each forehatch step is a box

If there are no steps to a forehatch it can be hard work climbing up, perhaps using a berth leeboard as one foothold, and maybe a locker door top as another. The steps shown here can be made at home, away from the boat, and secured in convenient places such as on a bulkhead, with one possibly on a berth front, and the top one on the inside of a cabin coaming, if it is vertical and the hatch edge is close.

Each step doubles as a small stowage locker with an open front. The sides of the mini locker are carried up to form fiddles so that when the boat is heeled or is rolling, a foot on the step will not slide off.

The bolts securing each step must be in line with the locker opening, otherwise drilling and countersinking the bolt holes will be difficult if not impossible.

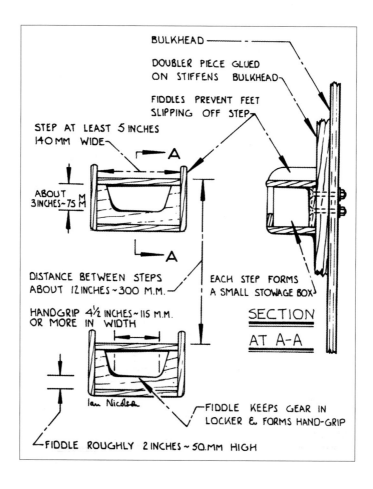

Forehatch ladder and emergency boat hook too

This forehatch ladder is light in weight, simple, and it takes up very little space. It is also easy to remove during the winter refit. For a boat that is permanently in a marina, and therefore seldom has any need for a boat hook, this ladder has a second use: it can, with just a little delay, be converted into the latter. For a bluewater cruiser it has the advantage of being a reserve boat hook, because these items of deck gear are more often lost overboard than worn out.

If the ladder is never to be used as a boat hook, it can be heavier and made out of steel. There should never be just one bolt at the top, and if the ladder is to be taken down in a hurry the nuts should either be of the butterfly type, or have lengths of rod welded to them so that they can be taken off without using tools.

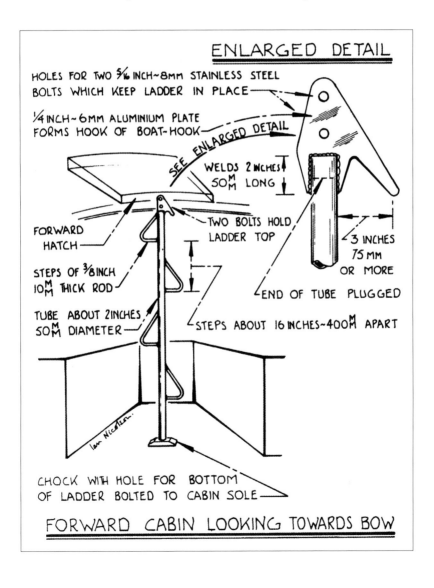

ENLARGED DETAIL

HOLES FOR TWO 5/16 INCH~8mm STAINLESS STEEL BOLTS WHICH KEEP LADDER IN PLACE

1/4 INCH~6MM ALUMINIUM PLATE FORMS HOOK OF BOAT-HOOK

SEE ENLARGED DETAIL

WELDS 2 INCHES 50mm LONG

FORWARD HATCH

TWO BOLTS HOLD LADDER TOP

3 INCHES 75 MM OR MORE

STEPS OF 3/8 INCH 10mm THICK ROD

END OF TUBE PLUGGED

TUBE ABOUT 2 INCHES 50mm DIAMETER

STEPS ABOUT 16 INCHES~400mm APART

CHOCK WITH HOLE FOR BOTTOM OF LADDER BOLTED TO CABIN SOLE

FORWARD CABIN LOOKING TOWARDS BOW

Side ladder I

Nothing looks smarter than a well-varnished hardwood side ladder, especially if the top hooks are of polished bronze. It used to be possible to buy special hooks that had double straps, one for the inside and one for the outside of each side piece of the ladder. Nowadays, it is usually necessary to make up the top hooks. The ones in this drawing are of 1-inch × ¼-inch (25-mm × 6-mm) metal flat bar, recessed into the wooden side piece. Instead of bending the metal straps over at the top, hardwood chocks are fitted, and as these lodge on the toe rail there is no risk that a wood or aluminium toe rail will be damaged.

A side ladder need not extend below the waterline, but if it does there is less chance a dinghy will get under it and lift it off; besides, if anyone goes bathing, or falls overboard, a ladder extending 18 inches (450 mm) below the water makes it much easier to get back aboard.

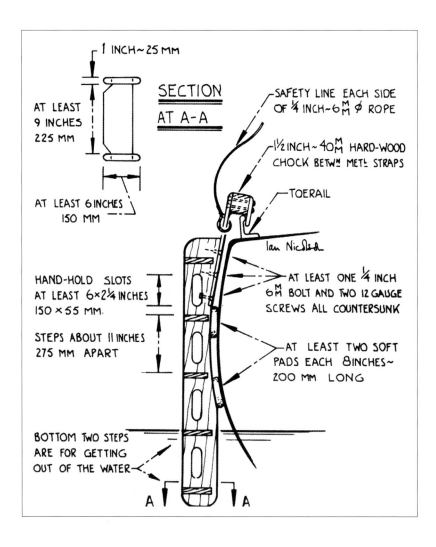

Side ladder II

The best side ladders are made from teak and glued up using epoxy resin. The area to be glued must be degreased, as the wood contains natural oils.

The steps can be morticed right through the sides, as shown in the middle drawing, but not every amateur feels confident he can cut a neat, tight mortice. Admittedly, with modern epoxy glues and their fillers a tight join is not essential, but for some amateurs the join shown at the bottom will be more popular. The fillet pieces are screwed and glued to the side pieces first. When the glue has set, the treads are screwed and glued in place. By locating the screws as shown, they will seldom be seen.

At the top of the top drawing the padding is shown. By keeping its fastenings on the sides of the ladder sides, there is no chance of scratching the topsides.

Anyone who wants a very light ladder can enlarge the hand slots in the sides and cut 1-inch (25-mm) slots in the treads. In any case, all edges should be well rounded.

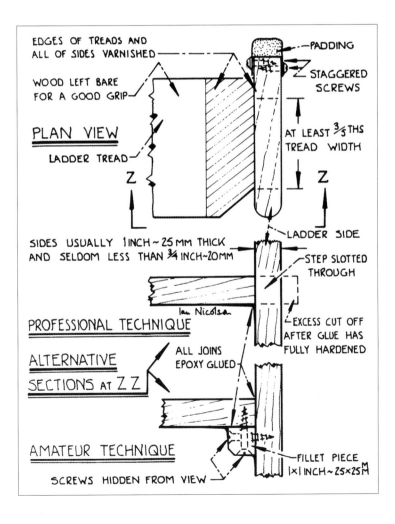

Props and ladder for boats laid up

This type of boat support is not cheap, especially when compared with the simple under-hull wooden shores made from lengths of untrimmed fir trees. On the other hand it does have a lot of advantages. It cannot come away from the boat by accident, even if the boat is being buffeted by hurricane force winds. It cannot be kicked away from under the boat by mistake or by vandals. It does not need moving when the underside is being anti-fouled, and it can be used to carry a concealed electric cable up to a light for scaring burglars as well as to a handy board for plug sockets.

Perhaps the biggest asset is the set of steps, which obviates the need to have a ladder. Admittedly, it is not easy to climb these sorts of steps, especially when carrying gear.

The tube should be about 1½ inches (35 mm) in diameter for every 10 feet (3 m) of boat length. It is best to have the steps extending up well above deck level, so that they act as hand holds.

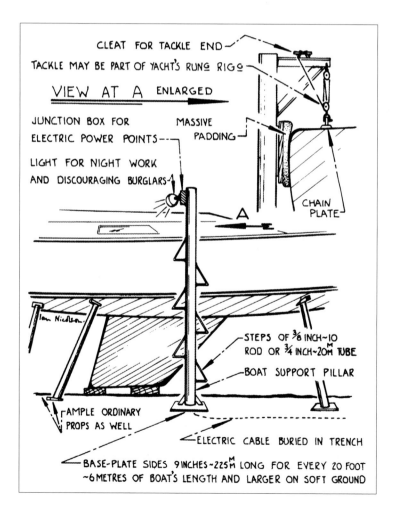

CHAPTER 13

Cockpit Comfort

Cockpit footrest

There is nothing as uncomfortable as sitting in a boat which is heeling, and not being able to put both feet on a secure resting place. The adjustable bar, extending fore-and-aft through the well of the cockpit is a well-known fitting, though it is not liked by everyone because when tacking there is a chance that someone will fall over it. But we cannot always cater for the heavy-footed people, and in time they learn to avoid these obstructions. In any case, for a series of short tacks the footrest can be removed and stowed out of the way.

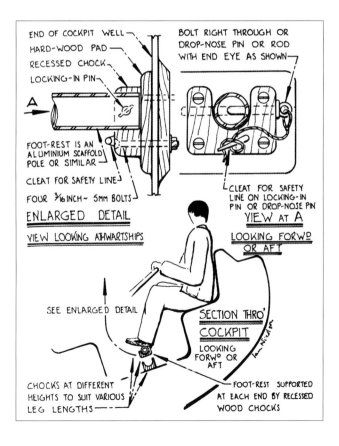

There should be at least two, and perhaps as many as four, different sets of chocks to cater for very small and very large people. A single, vertical hardwood backing pad can be used with a succession of slotted thick chocks to take each tube end.

If a short rod of stainless steel, perhaps made from a bolt with the head drilled and the thread ground away, is used to hold the tube in place, it should have a safety line secured tightly to its cleat to prevent it coming out.

Avoiding sore backs

When sitting outboard in a typical modern sailing yacht, the helmsman and crew find they are leaning against the lifelines. As these are normally thin wires, even tough, experienced crew object to the discomfort and the helmsman finds it hard to concentrate if his back is being 'sawn in half'.

A well-tried way of improving the cockpit comfort is to fit lengths of webbing instead of lifeline wires. Toestrap webbing can be bought from chandlers or from sailmakers. The latter will work eyes in each end, as shown at the top, but amateurs

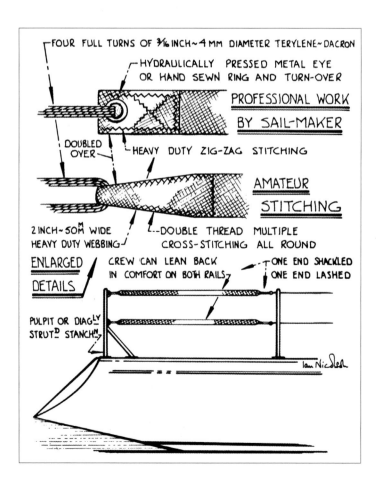

armed with nothing but some strong sail twine, a needle and palm and a little patience can stitch their own loops onto the webbing. If no metal eye is used, the webbing must be inspected monthly for chafe.

An end lashing or a rigging screw is essential to keep the webbing tight. Owners who are worried about windage, fold the tape edge to edge and stitch it together before putting the end eyes in place. A back-up wire, which may be of thin gauge, should run outboard of the webbing: this is compulsory under RORC safety rules.

Footrest combined with ladder

This helmsman's footrest can also be used as a ladder, either over the yacht's side, or up to a high quay. It is made from marine ply so that it is strong and will last. On each side, glued and screwed flanges of wood cover the ply edges and make the whole unit rigid. At each end are knee-shaped chocks which keep the footrest at the right height above the seats.

All edges must be well rounded and the whole fabrication should have four coats of varnish, or even more. People with large feet, and hence even larger sea boots, may prefer larger slots than those shown, and a good case can be made for assembling a rough mock-up or template of the whole unit, before starting to cut the wood.

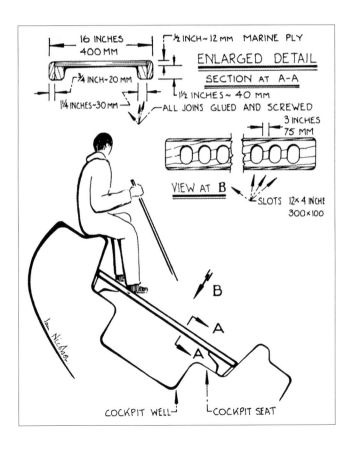

Adjustable footrest

When a boat is heeled, the helmsman needs a comfortable stance if he is to steer a good course and enjoy his job. Above all, he needs a footrest that he can adjust to his own height, and maybe change after an hour or two, to relax his muscles. He also needs a footrest that suits different angles of heel.

This footrest consists of two vertical pieces of wood shaped to suit the cockpit seats, and held together at each end by top and bottom wooden straps extending fore-and-aft. Inside each piece of wood there is a length of track. Two alternatives are shown here in the enlarged details at the top. On the left is a section of genoa sheet lead track with a tube about ¾ inch (20 mm) in diameter welded to each carriage on the two tracks. The plunges on the carriages locate the footrest bar.

The alternative, shown on the right, is an ordinary sail track with slides made up of either a steel T-bar section cut and filed to size, or alternatively a special short length of section welded up from steel flat bar. Wooden cross-pieces take the helmsman's feet and the carriages are fixed along the track by split pins or bolts through the tracks.

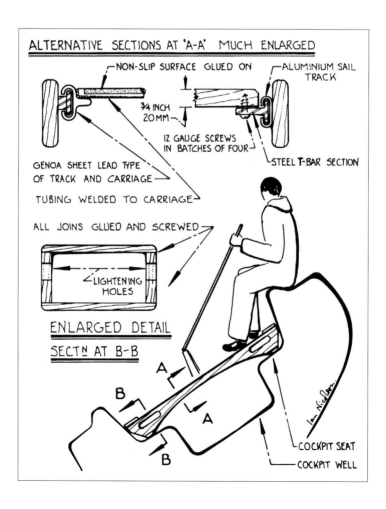

Cockpit to suit the tall and the short

Because people are not all the same height, they do not all favour the same width of cockpit well. What everyone does want is a bench which has the opposite seat edge just the right distance away for resting the feet on, when the boat is heeling.

The plan view shows a cockpit well which was originally too wide. It has been narrowed by a pair of tapering strakes of hardwood, so that it is narrow at the aft end and wider forward.

In practice, the tiller may obtrude, and also these ideal widths may be impossible to achieve without absurdly wide strakes. However, this sketch shows what to aim for. If in doubt, it is best to make the cockpit well slightly narrow, because the crew can always sit with knees well bent.

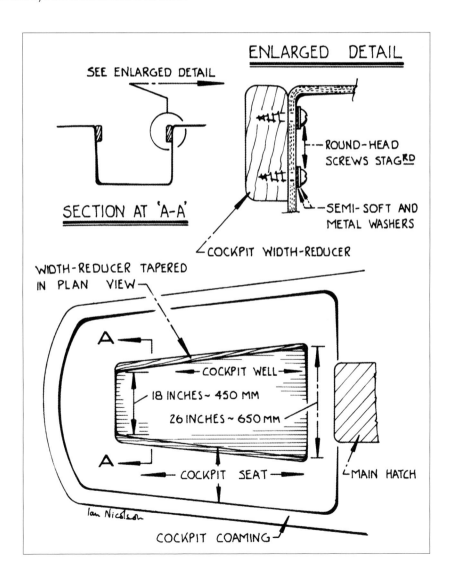

Ideas from Landamore's Oyster

Leslie Landamore built a 43-footer (13-m) Oyster type with this elegant helmsman's seat. It is shaped so that the helmsman sits level even when the yacht heels, a very desirable asset if the boat is steered with a wheel.

Under the ends of the seat there are lockers for gas cylinders, but because the seat is slatted, the lockers are difficult to detect, and this reduces the chances of the gas cylinders being stolen. Each locker lid needs a shock cord downhaul to prevent it flipping open in windy conditions, and, of course, a lock is needed.

The gas lockers drain into the cockpit, and the gas would accumulate in the well if the cockpit drains discharge below the waterline. The cockpit drains must also run downhill all the way, otherwise water will accumulate in the pipes and gas will not be able to escape. To save frequent swapping of gas bottles, they can be linked to the cooker in pairs or even in threes. However, when the boat is left for a period, or if there is to be no cooking for a day, all the bottles should be shut off, to eliminate any chance of leakages.

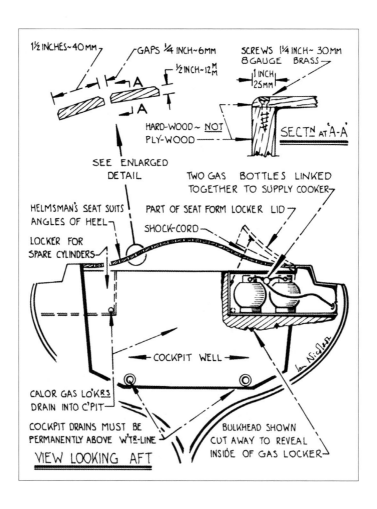

CHAPTER 14

Deck Fittings

Sunken, snug and safe

When building a new boat or changing an old one, there are a dozen arguments in favour of fitting a deep capacious forward pit. Working in it – sitting, standing or kneeling – the crew are far less likely to fall overboard than if up on deck. They can wedge themselves in, and should have some extra protection from wind and waves, especially if the toe-rail is extra high.

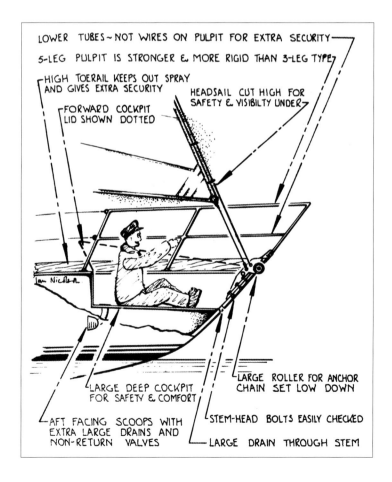

LOWER TUBES~NOT WIRES ON PULPIT FOR EXTRA SECURITY

5-LEG PULPIT IS STRONGER & MORE RIGID THAN 3-LEG TYPE

HIGH TOERAIL KEEPS OUT SPRAY AND GIVES EXTRA SECURITY

HEADSAIL CUT HIGH FOR SAFETY & VISIBILTY UNDER

FORWARD COCKPIT LID SHOWN DOTTED

LARGE ROLLER FOR ANCHOR CHAIN SET LOW DOWN

LARGE DEEP COCKPIT FOR SAFETY & COMFORT

STEM-HEAD BOLTS EASILY CHECKED

AFT FACING SCOOPS WITH EXTRA LARGE DRAINS AND NON-RETURN VALVES

LARGE DRAIN THROUGH STEM

Such a capacious pit must be well drained, hence the two aft and single bow outlets. Any two of these should be capable of emptying the pit in a couple of minutes, and in any case even a full pit should never be so big that it can contain enough weight of water to endanger the vessel. The bottom of the pit must be strong enough to support a full load of sea.

The bow roller is designed so that it is easy to use. Being lower and larger than usual, the crew have a less muscle-aching time heaving the anchor chain onto it, and over it.

When the foot of the headsail is higher than the pulpit, there will not only be a much better view forward and fewer bashed heads, but also the chafe on the sail will be reduced.

Anchor locker improvements

The fibreglass lids on many anchor lockers are slippery and none too strong. Both defects can be cured simultaneously by fixing hardwood strips on top. Teak is best but costly; afrormosia is tough and some people find it acceptable unvarnished. The tread strips should be fixed with close-spaced and well-staggered screws driven up from below. These wood battens form ridges that enable the crew to wedge their feet along the edge when the deck tilts.

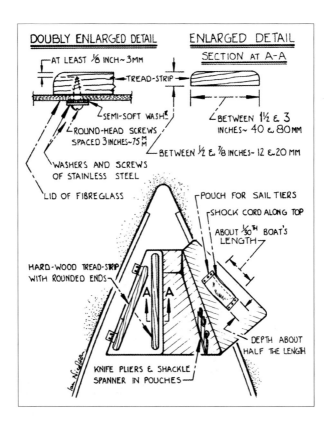

The inside of the locker is an ideal place to have a stowage pouch for sail tiers, and others for deck tools like a knife, a pair of pliers for jambed snap shackles, and a shackle spanner to open ordinary shackles. The inside of the locker may be padded with PVC or a similar cloth to prevent the ground tackle from damaging the fibreglass of the locker.

Noise deadeners

For anyone trying to sleep below, activity in the cockpit can be infuriating. In particular, if anyone bangs down a cockpit locker lid, the crack can waken even exhausted foredeck hands. The top right sketch shows how common domestic doorstops may be used to prevent lids banging down. The exact height of the rubber stop is critical, and is adjusted by planing away the top of the wood chock, before fitting the rubber knob.

Some lids can be quietened by putting simple rubber discs on the surface where the lid thumps down, but these may not stay in place for long. The bottom pair of sketches shows a rubber noise checker that is bolted in place so that when it wears out it is easy to replace, and until that time it will stay where it is put. The webbing handle has the virtue that it will cause no bruises.

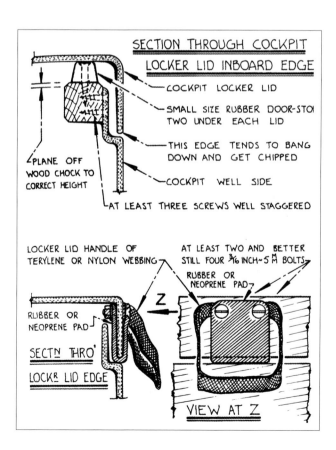

SECTION THROUGH COCKPIT LOCKER LID INBOARD EDGE

COCKPIT LOCKER LID

SMALL SIZE RUBBER DOOR-STOP TWO UNDER EACH LID

THIS EDGE TENDS TO BANG DOWN AND GET CHIPPED

PLANE OFF WOOD CHOCK TO CORRECT HEIGHT

COCKPIT WELL SIDE

AT LEAST THREE SCREWS WELL STAGGERED

LOCKER LID HANDLE OF TERYLENE OR NYLON WEBBING

AT LEAST TWO AND BETTER STILL FOUR $\frac{3}{16}$ INCH~5 M BOLTS

RUBBER OR NEOPRENE PAD

RUBBER OR NEOPRENE PAD

SECT$^{\underline{N}}$ THRO' LOCK$^{\underline{R}}$ LID EDGE

Z

VIEW AT Z

Bruise reducer

There is nothing as disconcerting as being bashed on the head by a boom. The gadget shown here minimizes damage in two ways:

1. The bright orange cover of the padding warns everyone in the firing line of the danger, and as it swings towards them, they have a chance to duck out of the way.
2. Even if the boom does make contact with someone's head, the ample padding changes a nasty thump into a more muted shove.

The padding may be of the same soft foam used in boat cushions, or the closed cell type which is made into life jackets. The latter does not soak up rain, as the cushion type will. These pads may be made up by anyone with a tough sewing machine, or by a local sailmaker.

When screwing the padding onto a boom, the fastenings near the end should be very close together, and all fastenings must be set on top of the boom, clear of the padding which covers the area likely to bang into a crew member.

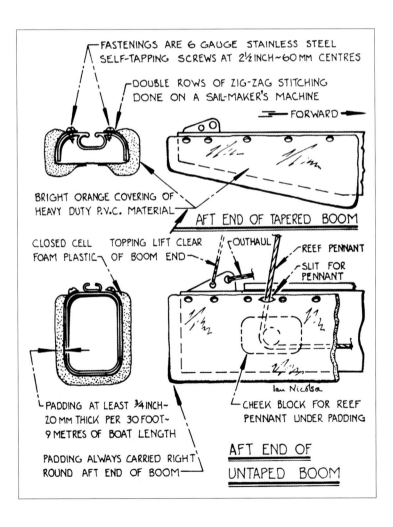

Rope tidy bags

Ropes in the cockpit tend to get snarled together. When the weather is bad, the crew may not have time to coil up those halyards and reefing lines that are led aft, or there may be seasickness aboard, so no one has the inclination to take sensible precautions.

Bags keep ropes separate, tidy and ready to run out quickly. Terylene (Dacron) or PVC or a similar tough cloth is used to make light, long-lasting bags. Each one should hold only one coil, and each bag should be slightly too large for its coil. Before ordering the bags from a local sailmaker, each coil should be measured to get the correct size. As a rough guide, a 40-foot (12-m) coil of ½-inch (12-mm) diameter rope needs a bag quite 18 inches (450 mm) deep and 10 inches (250 mm) across.

Sometimes, a length of shock cord is fitted inside the top outer seam to keep the bag closed.

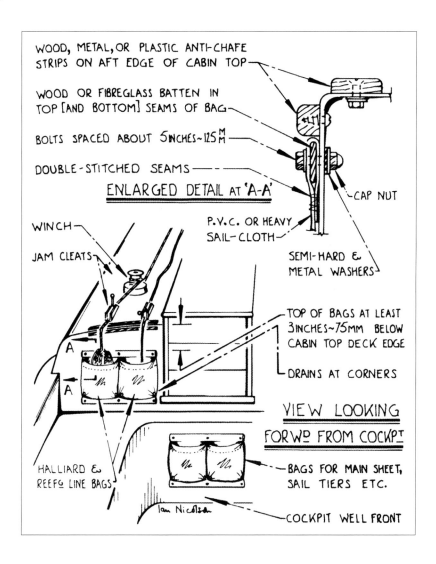

Three-legged mast pulpit or granny rail

For working at the bottom of the mast in anything but the quietest weather, there is nothing as helpful as a pulpit, sometimes called a 'granny rail'. This rather unkind expression disguises the fact that without the rail, work takes longer, people fall overboard more often, halyards are lost aloft more frequently and beginners are sometimes put off sailing.

The pulpit feet must be very strongly bolted down with at least three ¼-inch (6-mm) bolts in each flange. If the under-deck chocks are recessed for the nuts, there is less chance of a bloody scalp for tall crew members.

The legs may be fitted with ¼-inch (6-mm) diagonal stiffening bars at the top to make the whole fitting more rigid, and if two people are to use the pulpit in severe conditions, a four-legged pulpit would be better.

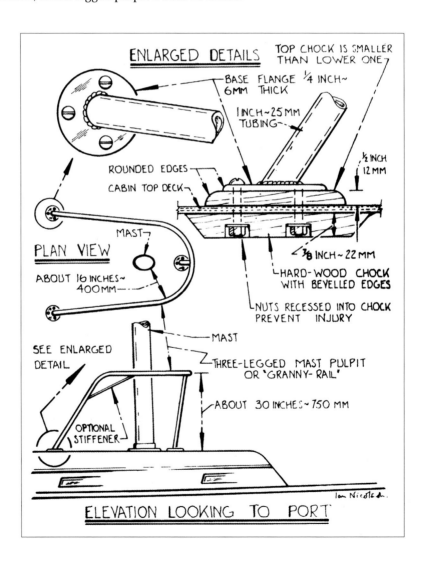

Granny rails which are also vents

Those rails, a bit like pulpits, which are located port and starboard of a mast, to hold the crew in position when halyards are being handled, or reefs put in, are often called 'granny rails'. They have to be high to be effective, and must be well bolted down if they are to survive harsh offshore conditions.

There is no reason why they should not double as ventilators, and being so high they should be acting in clear air, almost always above spray level. Even if spray does drive in, the slope of the top tubes should ensure that water will flow back out of the inlet hole. When the going gets too rough the air inlet is sealed by rotating the outer tube, shown in the enlarged detail at top right. This outer tube must be kept well lubricated, otherwise it may corrode onto the inner tube.

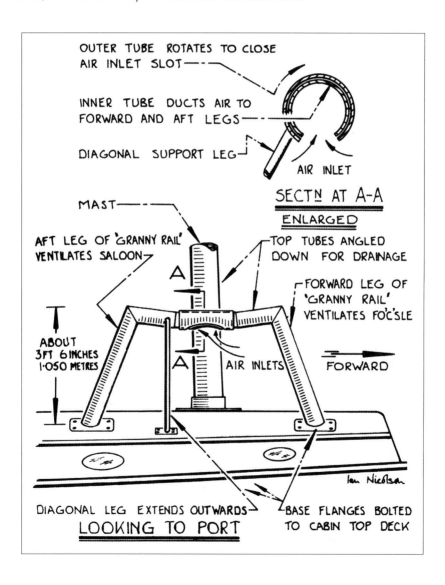

Wind sock ventilator

In hot climates, there is no more popular ventilator than the wind sock, because it channels masses of cooling air into the boat through the forehatch. These socks are made from spinnaker nylon because this is a hard wearing material which stows away in a small space. It is also wind tight so there is no loss of airflow through the pores, and it is widely available.

It is also soft and pliable, which means anyone can simply push aside the trunking of nylon cloth in the hatchway and climb in or out quite easily. The mouth of the sock must be kept open, and sail batten material is as convenient as anything for this. Some sailmakers just sew a length of fibreglass battening into the seam of the mouth, while others add a vertical or horizontal batten to hold the mouth in an oval shape, as shown here.

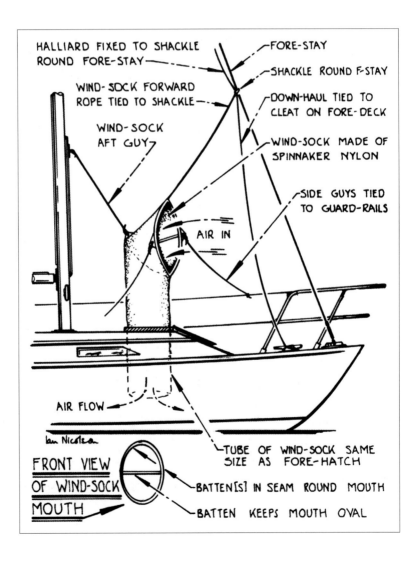

Mast steps

For long-range cruising, a set of steps up the mast may not be essential, but they do take a lot of the worry out of sailing. They make it so much more comfortable for the person going aloft. Even if he is in a bosun's chair, he will be able to steady himself all the way up, and he can take some of the load off the halyard.

The height of the steps depends on the owner's choice; he may feel that a jambed or unrove halyard is the most likely problem aloft, in which case he will have the steps located to deal with these problems. Alternatively, he may worry more about his masthead instruments, in which case he will arrange to have the steps fitted higher, to make working on the gadgetry easy.

Steps need a light line made fast all the way down, port and starboard, to prevent halyards from tangling with the steps.

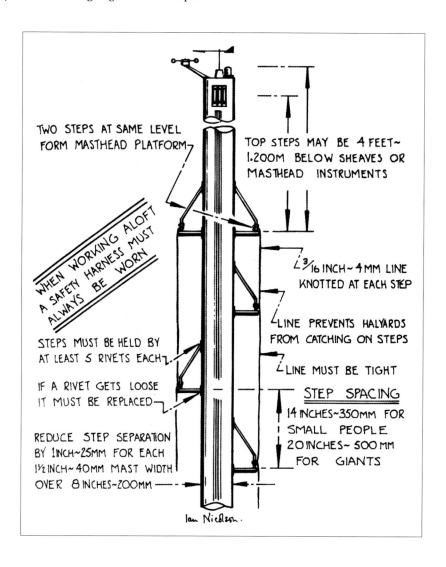

TWO STEPS AT SAME LEVEL FORM MASTHEAD PLATFORM

TOP STEPS MAY BE 4 FEET~ 1.200M BELOW SHEAVES OR MASTHEAD INSTRUMENTS

WHEN WORKING ALOFT A SAFETY HARNESS MUST ALWAYS BE WORN

³⁄₁₆ INCH~4MM LINE KNOTTED AT EACH STEP

LINE PREVENTS HALYARDS FROM CATCHING ON STEPS

STEPS MUST BE HELD BY AT LEAST 5 RIVETS EACH

LINE MUST BE TIGHT

IF A RIVET GETS LOOSE IT MUST BE REPLACED

STEP SPACING
14 INCHES~350MM FOR SMALL PEOPLE
20 INCHES~500 MM FOR GIANTS

REDUCE STEP SEPARATION BY 1 INCH~25MM FOR EACH 1½ INCH~40MM MAST WIDTH OVER 8 INCHES~200MM

Ian Nicolson.

No trouble on the foredeck with headsails

In strong winds, some headsails are hard to get down. A downhaul rove through extra-large eyelets at every other jib hank, or maybe every third or fourth, makes the job easy. If the sail fights hard, the downhaul can be taken on a winch on or near the mast.

During prolonged gales, headsail hanks sometimes wear through, and, even more often, come adrift. To avoid subsequent trouble when the sail is being lowered in a ferocious breeze, the eye on the halyard should have a shackle through it and round the forestay, so that the head of the sail is held close to the forestay. This shackle, if permanently used, also reduces the chance of the halyard being lost aloft. Even when it does get adrift, it is easier to recapture if it is shackled to the forestay.

By fitting another shackle to the tack pennant and round the forestay, a headsail is further kept under control when being hoisted or lowered in tough conditions. Also, the halyard end can easily be clipped onto the pennant when not in use.

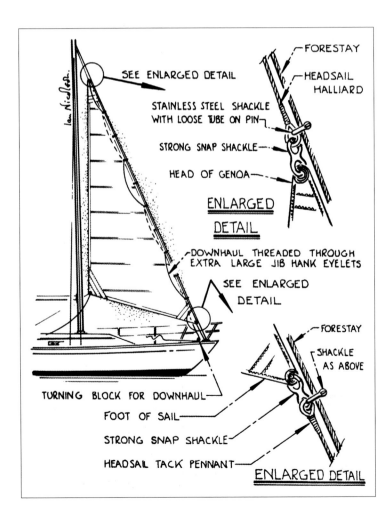

Anchor handling

To take the hard work out of lifting a heavy anchor on board, rig the spinnaker pole complete with its usual topping lift. A pair of light ropes steady the outer end of the pole, and an anchor lifting line is led from a mast winch via two turning blocks to the anchor. The boom is used like a derrick, and the halyard winches provide the power to lift the heavy weight. To bring the anchor inboard, once it has been hauled to the surface, the anchor lifting line is clipped on. The chain is now either disconnected, or slackened away, and the anchor hoisted to the boom end. The boom is swung aft a little, then topped up, and so the anchor is positioned over its chocks on deck. Lowering the anchor exactly where it is needed is easy enough as the lifting line is gently surged round the winch. The whole job can be done single-handed, slowly, stage by stage.

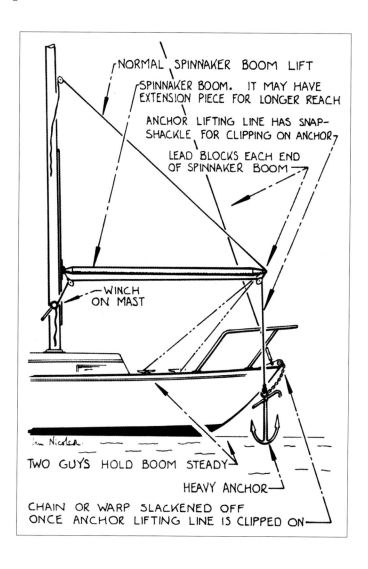

NORMAL SPINNAKER BOOM LIFT

SPINNAKER BOOM. IT MAY HAVE EXTENSION PIECE FOR LONGER REACH

ANCHOR LIFTING LINE HAS SNAP-SHACKLE FOR CLIPPING ON ANCHOR

LEAD BLOCKS EACH END OF SPINNAKER BOOM

WINCH ON MAST

TWO GUYS HOLD BOOM STEADY

HEAVY ANCHOR

CHAIN OR WARP SLACKENED OFF ONCE ANCHOR LIFTING LINE IS CLIPPED ON

Trip stopper

Any obstruction on deck can be dangerous, as well as hurtful, if the crew are inclined to trip over it. Those stainless steel tracks for the genoa sheet lead blocks are all too prone to stub toes, especially during a dark, wet night.

If tapered strips of wood are fixed along each side of the tracks, the foot is guided over the obstruction, and the track can be walked over quite comfortably. At each end of each track there must be a tapered chock to match the side strips. When this sort of end chock is fitted, the usual plastic end stops which prevent the track cars from coming off accidentally are not needed.

This drawing shows another gadget: the short length of plastic tube (or non-ferrous spring) round the shackle which holds the sheet lead block. The tube prevents the block from falling over and banging on deck. In light airs, when a sail is filling and collapsing, the constant chatter of a block on the deck will keep the watch below wide awake, which is maddening.

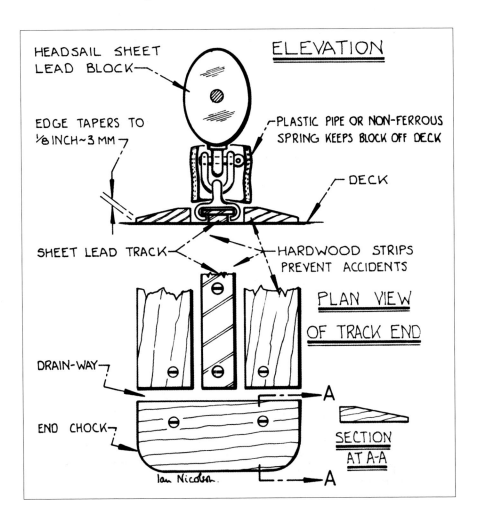

Reducing weather helm

The discomfort of a tugging tiller can be reduced by applying various techniques, even if the heavy weather helm cannot be fully eliminated.

The most obvious trick is to make the tiller extra long, so that the work needed to hold it to windward is less. Just as important, the tiller extension should be of the ladder type, which is easier to grasp for long periods, not the straight tube type that is used on dinghies.

Another technique is to increase the rudder area forward of the pivot line. This needs doing with caution, and as a general rule the total area forward of the centre of the stock should never exceed one sixth of the total area. Naval architects often advise that the area is increased progressively (on the leading edge), as shown in the enlarged detail.

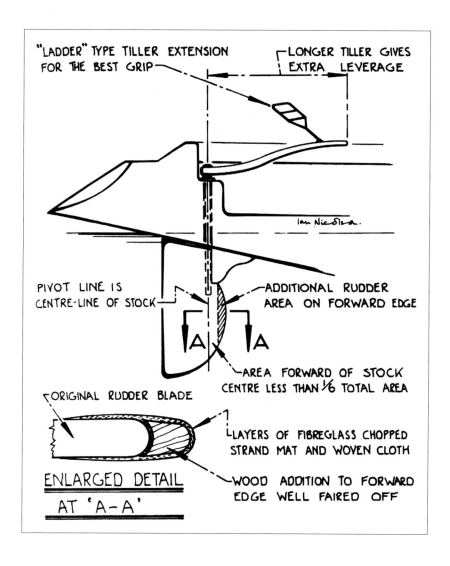

"LADDER" TYPE TILLER EXTENSION FOR THE BEST GRIP

LONGER TILLER GIVES EXTRA LEVERAGE

Ian Nicolson.

PIVOT LINE IS CENTRE-LINE OF STOCK

ADDITIONAL RUDDER AREA ON FORWARD EDGE

A A

AREA FORWARD OF STOCK CENTRE LESS THAN ⅙ TOTAL AREA

ORIGINAL RUDDER BLADE

LAYERS OF FIBREGLASS CHOPPED STRAND MAT AND WOVEN CLOTH

ENLARGED DETAIL AT 'A-A'

WOOD ADDITION TO FORWARD EDGE WELL FAIRED OFF

Tiller lock

There are various types of lock for holding the tiller in a particular position. Some consist of a form of ratchet, with a tongue on the bottom of the tiller which engages in the ratchet; some are in the form of a wooden plank athwartships under the tiller with holes in to take pegs, one peg going each side of the tiller. These locks are not entirely satisfactory because the spacing of the ratchet troughs, or the holes in the plank for the pegs, may not suit the required tiller location precisely.

The arrangement shown here is infinitely variable, allowing the helmsman to set the tiller just where he wants it to achieve self-steering, or make the boat heave-to.

The type and size of jam cleat will depend on the size of rope and boat. The rope size shown will suit boats up to about 40 feet (12 m). Each rope should have a knot in it so that the helm cannot be put over more than 35 degrees. This is the limit for effective steering, because beyond that angle water flow breaks down and steering becomes less effective. Going astern with more than 35 degrees of helm can cause damage to a rudder.

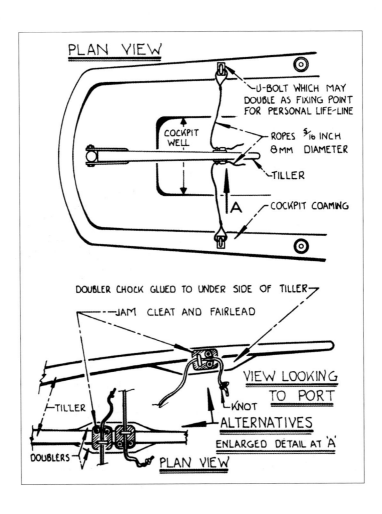

CHAPTER 15

Keeping the Spray at Bay

Sailing in the dry

This trick works well with old, steady yachts, but for very flighty craft, which will not hold their course even for a few seconds, it may be hard to operate. The helmsman sits in the warmth and comfort of the cabin and steers a course by compass. The hand-bearing compass, if big and steady enough, can be mounted on the chart table, or the main compass may be one of the semi-portable types with more than one mounting bracket. Care is needed to ensure that there is no steel near the compass which will affect it.

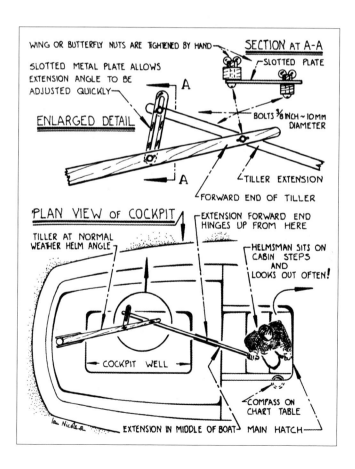

WING OR BUTTERFLY NUTS ARE TIGHTENED BY HAND

SLOTTED METAL PLATE ALLOWS EXTENSION ANGLE TO BE ADJUSTED QUICKLY

SECTION AT A-A

SLOTTED PLATE

ENLARGED DETAIL

BOLTS ⅜ INCH ~ 10MM DIAMETER

TILLER EXTENSION

FORWARD END OF TILLER

PLAN VIEW OF COCKPIT

EXTENSION FORWARD END HINGES UP FROM HERE

TILLER AT NORMAL WEATHER HELM ANGLE

HELMSMAN SITS ON CABIN STEPS AND LOOKS OUT OFTEN!

COCKPIT WELL

COMPASS ON CHART TABLE

EXTENSION IN MIDDLE OF BOAT

MAIN HATCH

The fore end of the tiller extension should pass through the cabin entrance at the middle, when the tiller is in the normal weather helm position if the boat is going to windward or reaching. Just occasionally, the helmsman may need to make a big movement of the extension, and so he will have to hinge its forward end up to clear the door posts of the cabin entrance.

To steer when below decks takes practice and skill, but in time it is possible to cook with one hand and steer with the other.

Cockpit protection

Quite a few yachts, especially motorsailers, are having windscreens fitted at the fore end of the cockpit. It can be a costly exercise, but if the window components are standard, or are made rectangular, or at least with two opposite sides parallel, the total expense can be limited.

It is almost always best to set the windscreen forward of the bulkhead at the fore end of the cockpit, and secure wood 'grounds' across the cabin top deck. The tops of these wood sections are straight, while the bottom faces follow the curve of the

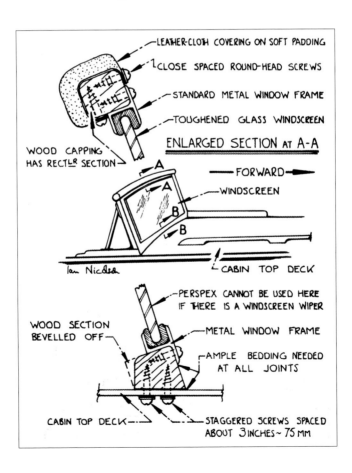

decking. Through bolts or ample screws driven upwards hold the grounds in place, as shown in the bottom detail.

The top enlarged section shows one good way of making the top edge strong and safe. The padding reduces the chance of injury, and if there is a cockpit hood this can be made watertight at the forward end by pulling it down tight onto the padding.

A better view, especially in thick weather

Nothing makes a boat as comfortable as an enclosed, or partly enclosed, steering position. However, in foggy or rainy conditions the windscreen becomes covered over with moisture, and it can be hard to see forward. A marvellous invention is the divided window, which has a top half hinged at the top. In foggy weather the upper part of the window is opened and the view much improved. If the fog turns to rain, the window is lowered till the helmsman just has a thin slit to look through, but no rain gets in, and not much draught either.

Naturally, the height of the helmsman's eye is critical, as it must be just above the top of the bottom part of the window, so that there is a good view over the sealing

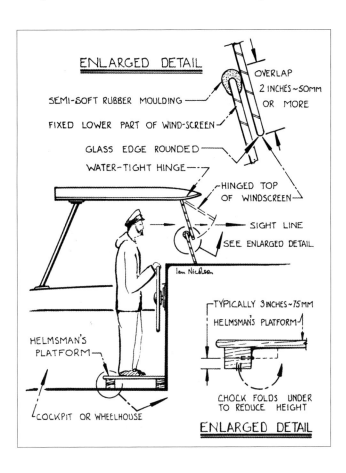

strip, shown in the enlarged detail (top right). At the bottom right the detail shows a simple way of making the helmsman's platform vary in height. There can be two folding chocks (at the front, and back naturally), instead of the single folding one sketched, to give more height variations.

A breakwater for dryness

Water pouring aft along a deck finds its way through hatch edges, into ventilators and generally works its way below. A good way to keep the aft deck and inside of a boat drier is to have a breakwater, or even two, on the foredeck.

A breakwater must be strong and it often needs metal support brackets if the height is over about 8 inches (200 mm). And if it is too high to step over easily, it needs one or two doorways through. The brackets are usually bolted to the deck and the breakwater, and are typically 3 feet (1 m) apart. They should extend to the end of the breakwater, which must reach to the deck edge for full effectiveness.

It is important that the breakwater looks smart since it stands out so boldly. The wood used must match the rest of the deck trim if the final finish is varnish, as it

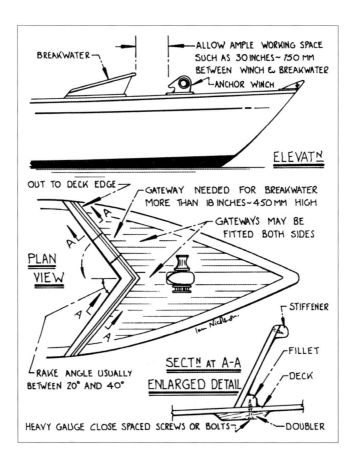

usually is. All edges must be very well rounded, and the bottom edge must be bedded in a semi-soft waterproof compound.

No water down the chain hole

A persistent trickle of water through the anchor chain hole when offshore is common on a wide variety of craft. It is surprisingly difficult to stop rain and spray getting into the hull through this access, because the hole is usually awkwardly hidden under parts of the anchor winch.

To get round the problem, a breakwater is fitted right round the winch. It is made as high as is reasonably convenient, especially forward where the waves come tumbling aboard. A tightly fitted cover goes over the top of the breakwater. This protects the winch from the weather and stops rain dribbling below. To make the cover fit snugly there must be a double lip on the top of the breakwater, and the lashing in the seam round the cover must be at least ¼ inch (6 mm) thick so that it can be pulled really tight.

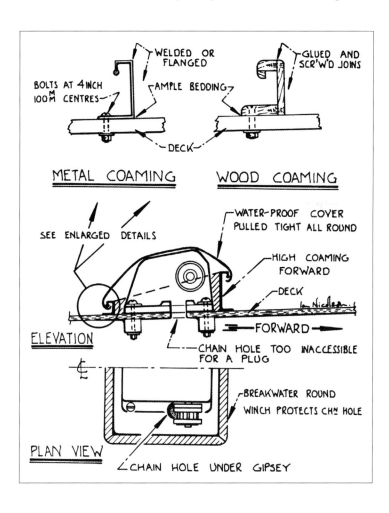

Watertight chain pipe

Pipes through foredecks for anchor chains seldom have good covers. Some have metal lids with slots that fit over one link of the chain, but in rough weather this lets in bucketfuls of water every hour. What is needed is a device that excludes all water, even if the boat goes upside down.

This simple device is made by taking a chock of a hardwood like teak, mahogany or iroko, and turning it or whittling it to fit the chain pipe. The plug must taper slightly so that it can be jammed in place, and at first it must be made a little too large. Next, it is sawn vertically and the exposed faces chipped away till the chain lies fairly snugly between the two halves. The plug is now glued together. The hole where the chain passes through is blocked at the bottom with plasticine, plastic, wood or even a stopping compound, and epoxy resin is poured down the hole at the top. This binds the chain and the plug together and ensures permanent watertightness.

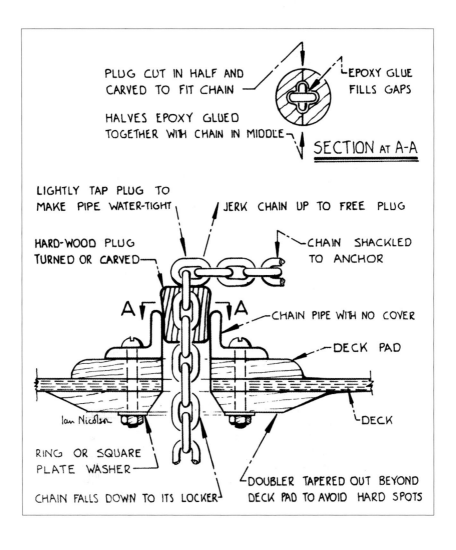

PLUG CUT IN HALF AND CARVED TO FIT CHAIN

EPOXY GLUE FILLS GAPS

HALVES EPOXY GLUED TOGETHER WITH CHAIN IN MIDDLE

SECTION AT A-A

LIGHTLY TAP PLUG TO MAKE PIPE WATER-TIGHT

JERK CHAIN UP TO FREE PLUG

HARD-WOOD PLUG TURNED OR CARVED

CHAIN SHACKLED TO ANCHOR

A

A

CHAIN PIPE WITH NO COVER

DECK PAD

Ian Nicolson

DECK

RING OR SQUARE PLATE WASHER

DOUBLER TAPERED OUT BEYOND DECK PAD TO AVOID HARD SPOTS

CHAIN FALLS DOWN TO ITS LOCKER

Electric cables through the deck

Down the mast there are often several cables that have to be led through the deck. On some boats the wires are taken to so-called waterproof plugs which tend to leak. On other craft the cables pass through glands that seldom keep the water out after two or three years' use.

This multiple gland is ingenious because it can be made to take any number of cables, and it is also cheap and simple to fabricate from easily available materials. It will stand up to heavy feet and heavy weather and if it is of varnished teak it will be very smart.

The outer two strips of wood are bolted down, and the middle one is hinged. Before it is shut down, all the cables are passed through their holes, then lots of soft bedding is squeezed around the wires where they go through the metal or plastic plate. When the lid is shut down, this forces the bedding tight round each cable, and the two (or more) turn-buttons keep the lid tight.

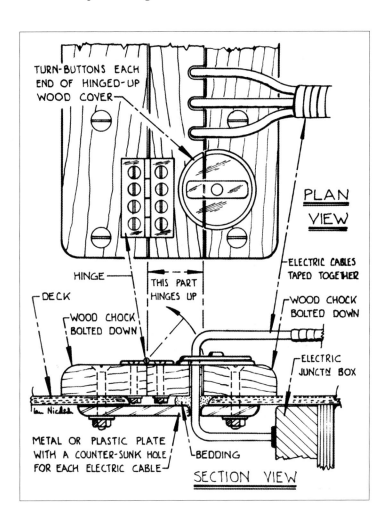

CHAPTER 16

Hatches and Cockpit Lockers

Drip-proof main hatch

Most main hatches let water in at the slides, and at the forward end. This design is in the form of an up-turned tray, and when shut the rubber seal along each side and across the front is bedded down tight on the coamings that stand up from the cabin top deck.

To open the hatch, the aft end is lifted enough for the aft turned-down lip to clear the side coamings, then the whole hatch is pushed forward. It is hinged to a sliding unit that glides forward along two tracks, which are set closer together than the hatch sides.

Each part of the hatch is easily made and simple to repair. It can be fitted on a wide variety of boats, the top can have a big window in it, and the design permits any level of massiveness or lightness, according to whether the boat is for long-range cruising or inshore racing.

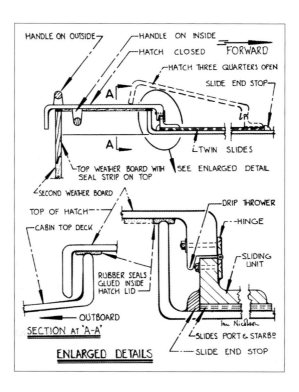

HANDLE ON OUTSIDE — HANDLE ON INSIDE
— HATCH CLOSED FORWARD
— HATCH THREE QUARTERS OPEN
SLIDE END STOP —
A
A
TWIN SLIDES
— TOP WEATHER BOARD WITH SEE ENLARGED DETAIL
SEAL STRIP ON TOP
— SECOND WEATHER BOARD
TOP OF HATCH — — DRIP THROWER
— CABIN TOP DECK — HINGE
SLIDING UNIT
RUBBER SEALS GLUED INSIDE HATCH LID
— OUTBOARD
Ian Nicolson
SECTION AT 'A-A'
— SLIDES PORT & STARB⁰
— SLIDE END STOP
ENLARGED DETAILS

Totally watertight low-cost hatch

This simple hatch can be made of aluminium or steel by any amateur or professional metalworker for very little money. The size will be made to suit the space on deck, the requirements of the crew (who may want to pass large sail bags through), and the structure of the boat.

The hatch height too can be adjusted to suit circumstances. If it is on the main weather deck and the boat is to go far offshore it will be best to have coamings 8 inches (200 mm) high or even more. But if it is on a cabin top, and the boat is a sleek one with no serious projections, then the coamings may be 2 inches (50 mm) high or even less.

If the hatch is of steel, the top and lower parts can be galvanised before being joined at the hinges. In addition, or instead of galvanising, the steel may be coated with epoxy paint.

The hasp and bolts or clips fitted will depend on what is available, or what can most easily be made up.

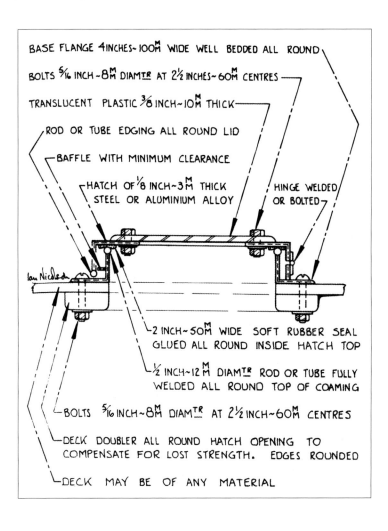

Watertightness is ensured by the wide, soft rubber strip that is compressed on the coaming top.

This is so much better than the narrow rubber used on so many commercial hatches, because the latter lack ample width of glued surface and the rubber sometimes works clear of the metal edge.

Water barriers on a sliding hatch

Driving spray filters in around the edges of sliding hatches unless precautions are taken. Shown here are some of the barriers, seals and baffles which can be fitted to many types of sliding hatch. On the left is a simple wood lip hinged onto the aft flange of the hatch top. When the hatch is closed, the lip folds down and shuts off that narrow gap above the top weatherboard.

To stop water driving in under the forward end of the hatch, there is the usual box or 'garage' over, but this is far from watertight. It can be made tighter by fitting

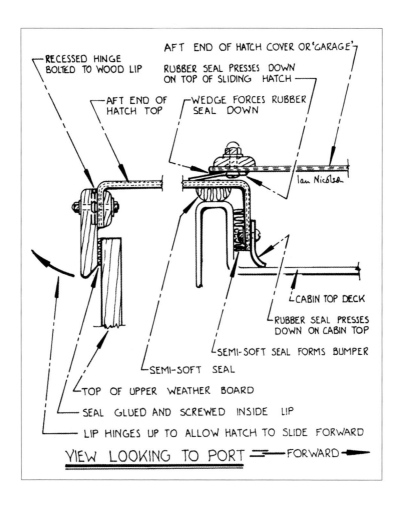

AFT END OF HATCH COVER OR 'GARAGE'

RECESSED HINGE BOLTED TO WOOD LIP

RUBBER SEAL PRESSES DOWN ON TOP OF SLIDING HATCH

AFT END OF HATCH TOP

WEDGE FORCES RUBBER SEAL DOWN

Ian Nicolson

CABIN TOP DECK

RUBBER SEAL PRESSES DOWN ON CABIN TOP

SEMI-SOFT SEAL FORMS BUMPER

SEMI-SOFT SEAL

TOP OF UPPER WEATHER BOARD

SEAL GLUED AND SCREWED INSIDE LIP

LIP HINGES UP TO ALLOW HATCH TO SLIDE FORWARD

VIEW LOOKING TO PORT ═══ FORWARD ➤

a rubber sealing piece under the aft edge of the cover, designed to press down constantly on the hatch top. As a second line of defence, there is that vertical seal bolted to the forward flange of the hatch top, and a third line of resistance is the semi-soft seal that doubles as a bumper and stops the hatch banging into the vertical flange at the aft end of the cabin top deck. Just for good measure there is another similar seal above. Naturally, it is not always possible or necessary to have all these seals on one boat.

Main hatch protection

The usual waterproof cloth hood over a main hatch needs renewing every few years. Also, it is not strong, and in very severe conditions it may be swept away. The one shown here is much tougher, being made from aluminium alloy plate. It has thick plastic windows bolted at 2-inch (50-mm) intervals all round, so that the joins round the window edges form thick bands of stiffening. More reinforcing is provided by the handrails and the flange round the aft edge, which is in the form of a gutter on edge. This prevents water flowing aft and under the cover.

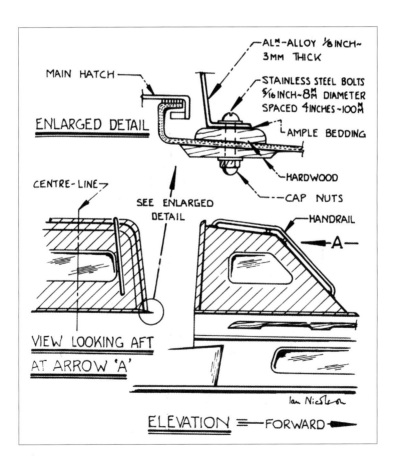

This sort of hatch hood can be made semi-permanent, so it is left off the boat when she is gently weekending inshore. Then, when she goes far from home into deep waters, the hood is bolted in place. It would suit a boat which, for instance, does a limited number of rough races each year, and perhaps one long cruise. For these events she would have the hood bolted in position, but for the rest of the time it would be left at home in the yacht's store.

Leak-free sliding hatch

Not only is this hatch unusually well guarded against water seepage, it is also low in profile. For a racing boat where the crew want to be able to walk unimpeded over the hatch, this type is almost as good as one which is entirely flush with the deck. The latter type is likely to be expensive and in practice less than completely watertight.

The top section shows how the hatch sides and aft end are tightly clamped down, forcing the Perspex top to lodge right down on the slides. In practice, the Perspex may wear too fast if the hatch is often used, so mating strips of metal would, in this

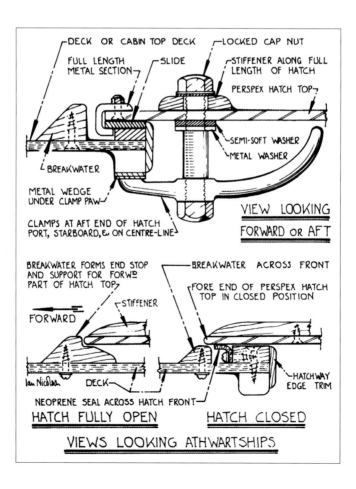

case, be secured on the underside of the Perspex to run on the slides. Alternatively, metal or a suitable hard-wearing plastic can be used instead of Perspex.

The bottom sketches show how the fore-end of the cover is supported and water prevented from driving in, these details can be built into other designs of hatch.

Wood forehatch with improvements

This hatch is a traditional design with improvements. Water that seeps in between the cover and the outer coaming drains out again via the drain holes. These are ⅝ inch (15 mm) in diameter or more so that they do not get bunged up with debris.

Around the outside at the bottom there is a fillet moulding, but this is only an extra line of defence because those upward driven screws, zigzagged at 3-inch (75-mm) intervals, not only secure the hatch to the deck tightly but also keep out leaks.

By having the Perspex right to the edge of the hatch there is less maintenance, and the traditional type of edge moulding is eliminated. This again reduces chances of leaks in future years, when the hatch is beginning to weather. Non-slip strips have to be close together so that there is no chance of a shoe getting between them, and so skidding.

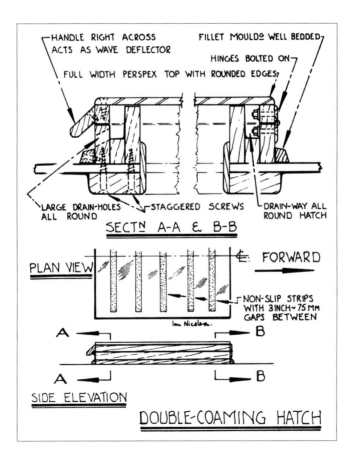

A wooden handle extends right across the aft edge and deflects solid water from the top of the outer coaming. This coaming, as well as the inner one and the filler piece between, are all the same thickness and are glued together.

Changing a forehatch

The top sketch shows typical defects in an inefficient hatch. Water can easily run in under the outer flange and the rubber seal may be right away from the upstanding flange or not compressed tightly enough, or the rubber seal may be discontinuous, or it may have perished. Often the cheapest way of getting a watertight hatch is to buy a standard Lewmar or similar hatch and bolt it down over the deck opening.

It is normally necessary to build up some form of coaming to take the new hatch, and in the lower sketch here, a wood coaming and underdeck doubler is shown. However, for anybody who prefers to work in fibreglass or metal, the new base for the hatch can be in these materials.

Whatever form the new hatch base takes, there must be very good seals wherever water is going to try to penetrate. The cost of the bedding compound is nothing

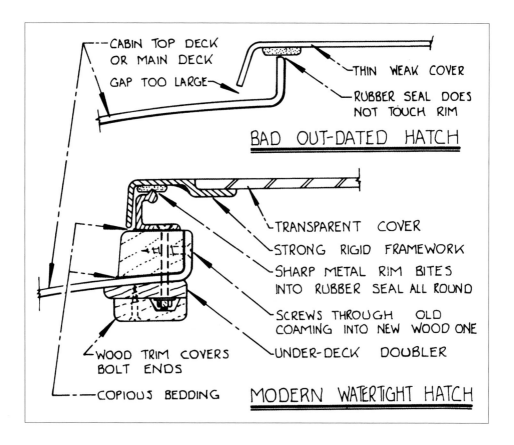

CABIN TOP DECK OR MAIN DECK

GAP TOO LARGE

THIN WEAK COVER

RUBBER SEAL DOES NOT TOUCH RIM

BAD OUT-DATED HATCH

TRANSPARENT COVER

STRONG RIGID FRAMEWORK

SHARP METAL RIM BITES INTO RUBBER SEAL ALL ROUND

SCREWS THROUGH OLD COAMING INTO NEW WOOD ONE

WOOD TRIM COVERS BOLT ENDS

UNDER-DECK DOUBLER

COPIOUS BEDDING

MODERN WATERTIGHT HATCH

compared with the discomfort of just one leak. This is why a hatch must be bolted, never screwed down. The bolts must be tightened progressively in the same way as an engine cylinder head is tightened, to avoid distortion. First of all, the nuts are wound up finger-tight, then the forward starboard one is slightly tightened with a spanner followed by the aft port one, then aft starboard, then forward port. Next, the intermediate bolts are gently spannered in the same order. Finally, all the nuts are pulled hard up in the same order.

Historical hatch

Myth of Malham, designed by Laurent Giles, had probably more influence on offshore sailing than any other yacht. This sketch shows her forehatch, which like every other component on board was carefully planned to do its job well.

The long-tailed hinges with widespread fastenings ensure that the hatch does not come adrift the first time someone carelessly stands on one side of the open lid. A combination of rubber seal and clamping bolts is common these days, but was not usual when this hatch was designed. No other arrangement is as reliable when it comes to excluding water from the inside of a yacht.

The rounded edges reduce the chances of ropes catching, look professional, and are easy to keep varnished or painted.

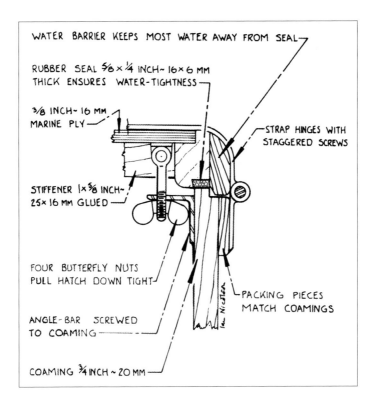

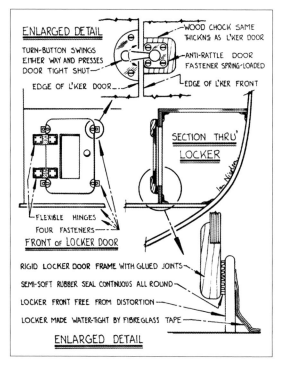

ENLARGED DETAIL

TURN-BUTTON SWINGS
EITHER WAY AND PRESSES
DOOR TIGHT SHUT

EDGE OF L'KER DOOR

WOOD CHOCK SAME
THICKNS AS L'KER DOOR

ANTI-RATTLE DOOR
FASTENER SPRING-LOADED

EDGE OF L'KER FRONT

SECTION THRU'
LOCKER

FLEXIBLE HINGES
FOUR FASTENERS
FRONT OF LOCKER DOOR

RIGID LOCKER DOOR FRAME WITH GLUED JOINTS
SEMI-SOFT RUBBER SEAL CONTINUOUS ALL ROUND
LOCKER FRONT FREE FROM DISTORTION
LOCKER MADE WATER-TIGHT BY FIBREGLASS TAPE

ENLARGED DETAIL

Watertight locker

Wet gear makes for a miserable crew, and one way to keep things dry is to stow them in a locker which has been made watertight. The inside joins are cleaned, prepared, then glassed over using three runs of 1½-oz fibreglass cloth and plenty of resin.

The door must be strong, rigid and watertight, with a wide semi-soft rubber seal glued to its back. This seal needs proper mitred corners and ample glue to hold it firmly all round. To allow the door to be forced tight on the locker front, there must be flexible hinges, which can be made of webbing with large plate washers over.

The door clips must press firmly against the door front, and they may be those anti-rattle door fasteners sold through chandlers. They are like strong turn-buttons and are easy to use in bad weather when the boat is jumping about.

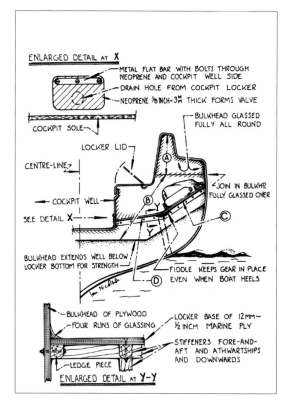

ENLARGED DETAIL AT X

METAL FLAT BAR WITH BOLTS THROUGH
NEOPRENE AND COCKPIT WELL SIDE

DRAIN HOLE FROM COCKPIT LOCKER

NEOPRENE ⅛ INCH-3ᵐ THICK FORMS VALVE

COCKPIT SOLE

BULKHEAD GLASSED
FULLY ALL ROUND

LOCKER LID

CENTRE-LINE

COCKPIT WELL

SEE DETAIL X

JOIN IN BULKHD
FULLY GLASSED OVER

BULKHEAD EXTENDS WELL BELOW
LOCKER BOTTOM FOR STRENGTH

FIDDLE KEEPS GEAR IN PLACE
EVEN WHEN BOAT HEELS

BULKHEAD OF PLYWOOD
FOUR RUNS OF GLASSING

LEDGE PIECE

LOCKER BASE OF 12MM~
½ INCH MARINE PLY

STIFFENERS FORE-AND-
AFT AND ATHWARTSHIPS
AND DOWNWARDS

ENLARGED DETAIL AT Y-Y

Making a cockpit locker watertight

Water gets into a boat via cockpit lockers unless they are designed to be watertight and drain overboard or into the cockpit well. To seal off a cockpit locker, end bulkheads of ply, probably ½ inch (12 mm) and certainly marine grade, are fitted. Where the shape is awkward, the top part will probably be fitted first, then the lower part, which should extend well down. In the sketch, the bottom section goes right under the cockpit and so stiffens the well. When these two parts, A and B, have been fitted the locker bottom is fitted, parts C then D. The outboard part is tilted to encourage drainage and prevent the locker being too big and therefore a danger to the vessel in the event of flooding of the cockpit.

As the crew will sometimes want to climb into the locker, and there may at other times be a weight of water in it, the bottom needs proper reinforcing, as shown bottom left. Parts C and D are glassed together and glassed around the edge. The bolts for many boats will be about ⅜ inch (8 mm) in diameter, spaced typically about 8 inches (200 mm).

A simple pair of valves, made from Neoprene sheeting, ensures that water drains out of the locker but does not trickle in from the cockpit well. (Detail top left.)

All this may seem a lot of work to prevent water getting inside the hull, and it is. But offshore sailors agree it is worth it. The alternative is to seal the cockpit locker lids and loose on-deck stowage. Some people prefer this approach.

Cockpit stowage

Underneath cockpit seats there are fine, large lockers. Their big capacity is so useful for storing fenders, sail bags and other bulky things, but it is a disadvantage when it comes to finding a home for a couple of sheet winch handles, or small lengths of rope, sail tiers, spare shackles and such-like.

A sailmaker will stitch up panniers to any given size. He may use heavy gauge sail cloth or PVC or one of the man-made modern substitutes for traditional canvas. If there are several pouches in one locker, it is a good idea to have them of different colours, then a stranger on board can be told that flares are in the red pouch, spinnaker sheets in the white one, and biscuits for the crew on watch in the blue one.

It may be advisable to have drain holes at the bottom of each pouch and these should be at least ½ inch (12 mm) in diameter to prevent dirt clogging them. Get the sailmaker to use zigzag stitching for ample strength along all the seams.

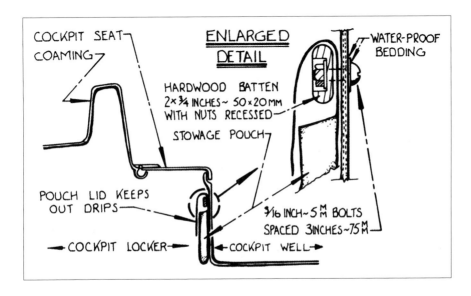

For access below the cockpit

Some cockpit wells have hatches in them, and these often leak. Some cockpits need new hatches through to the bottom, perhaps to reach a stern gland, or get a new tank in, or an old tank out. The hatch in the bottom of the well needs to be strong and properly sealed. This sketch shows a double seal for good measure, and an arrangement which can be used on a newly made or existing hatch.

Plywood, metal or fibreglass can be used for the hatch top. The holes for the vertical bolts round the hatch top are drilled under-size with the hatch lid positioned exactly, then the lid is taken off, and the holes in the angle bar carefully threaded.

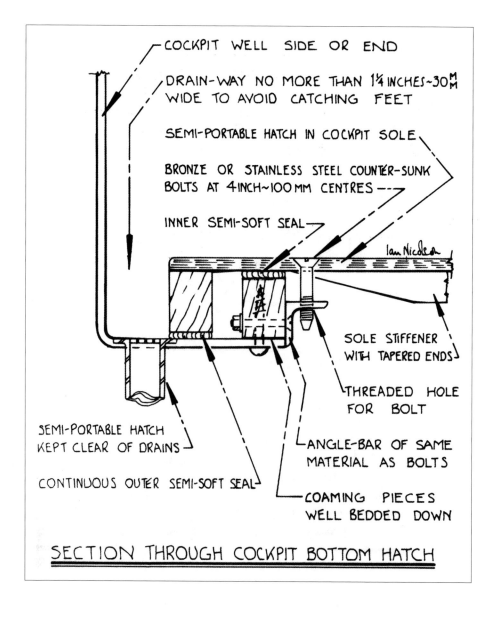

COCKPIT WELL SIDE OR END

DRAIN-WAY NO MORE THAN 1¼ INCHES~30 ᴹ WIDE TO AVOID CATCHING FEET

SEMI-PORTABLE HATCH IN COCKPIT SOLE

BRONZE OR STAINLESS STEEL COUNTER-SUNK BOLTS AT 4 INCH~100 MM CENTRES --→

INNER SEMI-SOFT SEAL →

Ian Nicolson

SOLE STIFFENER WITH TAPERED ENDS

THREADED HOLE FOR BOLT

SEMI-PORTABLE HATCH KEPT CLEAR OF DRAINS

ANGLE-BAR OF SAME MATERIAL AS BOLTS

CONTINUOUS OUTER SEMI-SOFT SEAL

COAMING PIECES WELL BEDDED DOWN

SECTION THROUGH COCKPIT BOTTOM HATCH

Boom Gallows

Goal post gallows

The basis of this boom gallows is a single length of steel or aluminium tube bent into the shape of a goal, with stiffening brackets at each top corner. In practice, there may be a forward- or aft-reaching diagonal leg each side to give the gallows stability in the fore-and-aft plane. These support struts, which are not shown in the drawing, can be quite short, and may be of the same thin bar or tube used to strengthen the corners.

The chock that supports the boom hangs upside down when not in use. When the boom is to be lodged in the gallows, the crew can locate the chock anywhere along the

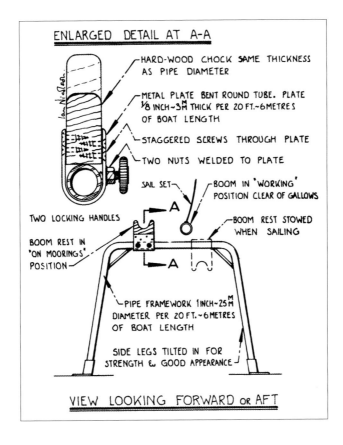

ENLARGED DETAIL AT A-A

HARD-WOOD CHOCK SAME THICKNESS AS PIPE DIAMETER

METAL PLATE BENT ROUND TUBE. PLATE ⅛ INCH~3ᴹ THICK PER 20 FT.~6 METRES OF BOAT LENGTH

STAGGERED SCREWS THROUGH PLATE

TWO NUTS WELDED TO PLATE

SAIL SET

BOOM IN 'WORKING' POSITION CLEAR OF GALLOWS

TWO LOCKING HANDLES

BOOM REST STOWED WHEN SAILING

BOOM REST IN 'ON MOORINGS' POSITION

PIPE FRAMEWORK 1 INCH~25ᴹ DIAMETER PER 20 FT.~6 METRES OF BOAT LENGTH

SIDE LEGS TILTED IN FOR STRENGTH & GOOD APPEARANCE

VIEW LOOKING FORWARD OR AFT

top length of piping. The two locking handles are turned tightly to secure the chock, and the boom dropped in place. When the chock is hanging below the pipe, the boom with sail set just clears the top of the whole affair.

This type of gallows makes a marvellous grab rail, and it can support a canvas dodger across the aft end of the cockpit, to make it snugger and safer.

Folding boom gallows

This simple gallows can be made to fold forward or aft, and by withdrawing both drop-nose pins each side, the whole fitting can be taken off the boat for storage or repairs.

Each base plate is secured to the deck or cabin top with four bolts. The height and width of the gallows is measured on the ship, and the tubular part is quickly made by anyone with a pipe-bending rig. The tube diameter would be about 1 inch (25 mm) in diameter for each 20 feet (6 m) of boat length. The diagonal stiffeners may not be essential, but they may prevent the gallows from being damaged by a careless gybe, if the mainsheet snags the gallows. The rod stiffeners will typically be one fifth of the tube diameter, or may be of tube one third the main tube diameter.

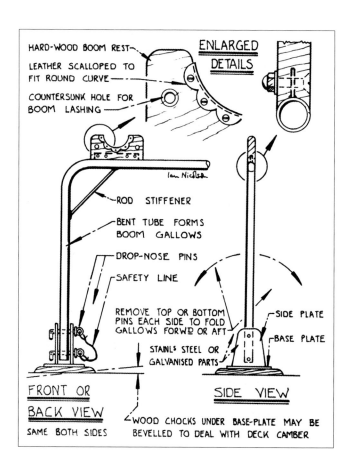

The top left sketch shows how the leather padding, designed to prevent the boom being damaged, is cut in scallops round the edge so that it will bend into a boom recess. Top right shows the metal plate welded to the top of the tube, to take the wooden boom rest.

Vertical sliding boom gallows

This device used to be common on every sort of yacht, but is now seen mainly on large ones. However, it does make life aboard more comfortable on many occasions. It forms a marvellous support if a boom tent is rigged; it can help when reefing because the aft end of the boom can be lashed into the gallows; it is ideal when the mainsail is lowered in severe weather and it forms a great grab rail. Of course, its main use is for when the boat is on moorings.

The one shown is based on the *Myth of Malham* type, which drops into the deck when out of use. Alternative details are shown. For instance, one side has a weight to keep the diagonal crosswire tight, the other a length of thick shock cord. One side has a raised bearing for the support tube, the other a sunken one. One side has a strong

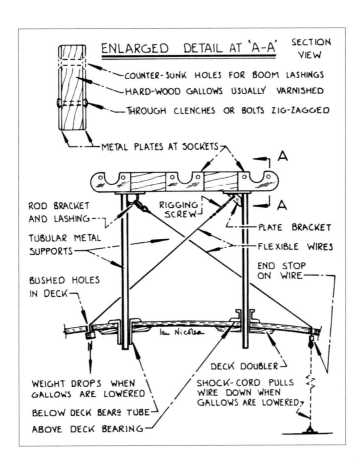

plate bracket at the top of the tube, the other a lighter, more elegant but thinner bar bracket. Owners can choose which style they favour for these parts.

Some people do not cut a central slot for the boom, since they feel one port and one starboard are adequate. Just a few owners use brass or bronze for all the metal parts and keep them polished which looks superb.

Y-type boom gallows

This type of gallows is compact, easy to stow and to erect in harbour, and it does not take up much deck space. Against that, it is awkward to put up at sea, and not much good for holding the boom in very severe conditions offshore unless a pair of guys are fitted to give steadiness athwartships.

It is handy if the boom is used to support a cockpit tent or awning.

The distance from the top to the metal strap should not be more than two and a half times the distance from the strap to the bottom. The bottom may be secured by another metal strap, or the upstanding peg shown here which fits firmly inside the end of the tube.

Traditionally, the top fork is padded with thick leather but as this material is expensive and not always easy to get, plastic pipe or PVC cloth may be used instead.

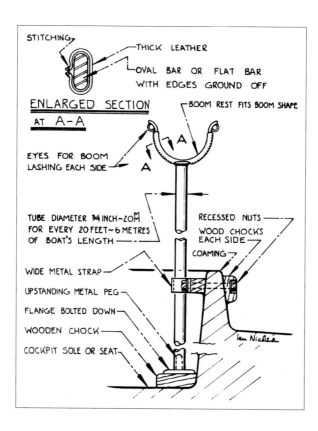

Wood single-leg boom gallows

This type of gallows is cheap to make, easy to stow, and suits those people who have woodworking facilities but no metal-working tools. The usual padding in the mouth of the Y has been omitted in favour of the easily available and simple-to-fit lashing of thin rope.

The arms of the Y are glued and held by metal fastenings which, theoretically, are superfluous, and could be removed when the glue has set hard. Alternative fastening techniques are shown on the left and right sides. Long copper clenches are not often seen these days but they do look nautical, sensible and, to most people, decorative.

The strap round the middle could be made from metal, but the arrangement shown has the virtue that it is built up from the same material of the same thickness as the rest of the gallows. The bottom socket too can be glued up from layers of the same wood. It should have a drain hole quite ⅝ inch (15 mm) in diameter to allow rain and spray to run away.

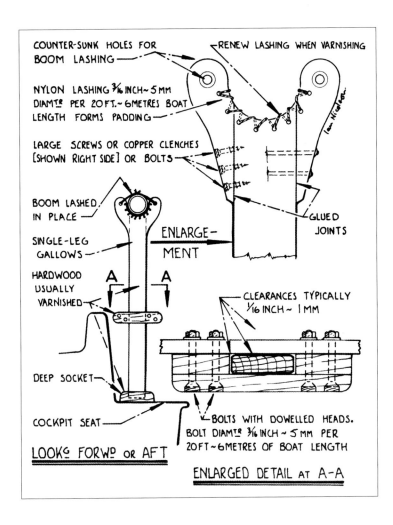

COUNTER-SUNK HOLES FOR BOOM LASHING

RENEW LASHING WHEN VARNISHING

NYLON LASHING ¾₆ INCH ~ 5 MM DIAM⁺ˢ PER 20 FT. ~ 6 METRES BOAT LENGTH FORMS PADDING

LARGE SCREWS OR COPPER CLENCHES [SHOWN RIGHT SIDE] OR BOLTS

BOOM LASHED IN PLACE

SINGLE-LEG GALLOWS

ENLARGEMENT

GLUED JOINTS

HARDWOOD USUALLY VARNISHED

CLEARANCES TYPICALLY ¹⁄₁₆ INCH ~ 1 MM

DEEP SOCKET

COCKPIT SEAT

LOOKᴳ FORWᴰ OR AFT

BOLTS WITH DOWELLED HEADS. BOLT DIAM⁺ˢ ³⁄₁₆ INCH ~ 5 MM PER 20 FT ~ 6 METRES OF BOAT LENGTH

ENLARGED DETAIL AT A-A

CHAPTER 18

Yacht's Tenders

Getting out to the yacht

The usual yacht's tender, whether it is an inflatable or a solid dinghy, is far from ideal for getting the crew and their gear out to the parent yacht at the beginning of a weekend. The popular type of tender is small so as to be easy to lift aboard and stow conveniently. It rows badly (especially if it is an inflatable), it does not carry much gear, sometimes not all the crew in one load, and it does not keep crew or their gear dry.

What so many people need is a larger rowing or outboard dinghy just for getting to and from the home moorings. It can be an old boat and it need not be light, provided

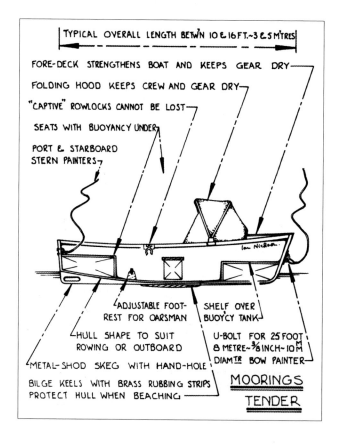

there are good facilities for launching it. Alternatively, it can be kept afloat all the time, possibly on a running line secured to the shore.

It does need to have ample freeboard, but if it is narrow it will be easier to row. It should be as large as possible and must support everyone on board with 6 inches (150 mm) of freeboard if flooded. It must be able to stand up to beaching, and have lots of fendering all round. Bow and stern painters are essential for tying alongside the parent yacht.

A more comfortable inflatable dinghy

Many inflatable dinghies have rowing thwarts that are inflatable, but such seats do not make for easy rowing. Those dinghies that have solid seats often have them set on top of the inflated side tubes, and this is too high for rowing. The top sketch shows how a rigid seat of the correct height is easily made.

Almost as important, there should be a stretcher or footrest for the oarsman. Since people are not all the same height, the footrest should be adjustable fore and aft. The bottom sketch is an enlargement of the bottom boards and footrest, as well as the footrest support.

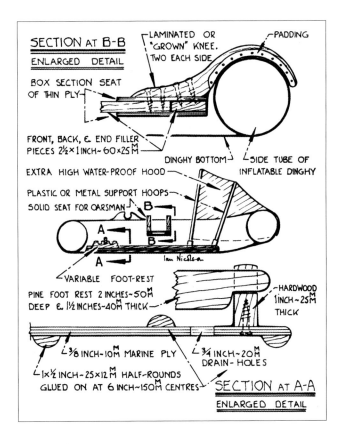

If an inflatable dinghy has no solid boards it will be hard to row. As marine ply is expensive and heavy, the bottom boards shown here are made up of a relatively thin sheet of ply with stiffeners top and bottom. The top ones act as toerails and prevent the crew slipping, the bottom ones reduce local chafe cause by the bottom boards.

Improving an inflatable

Most inflatables are hard to row, not least because they have no stretcher or footrest for the oarsman. It is difficult to fix a footrest in an inflatable that has no bottom boards, so the design show here, or a similar one, has to be used.

Wear on the bottom, especially aft, is the main reason why many inflatables leak and chafe to destruction. This trouble can be eliminated by doubling the bottom with a piece of canvas, tough rubberised cloth or a similar resilient material.

Even their best friends admit that it is a slow and tedious job blowing up an inflatable. Also, it can be frustrating trying to get the air out when the time comes to stow the dinghy in a small locker. A 12-volt air pump is now popular for inflating and deflating, both in emergencies and for everyday use.

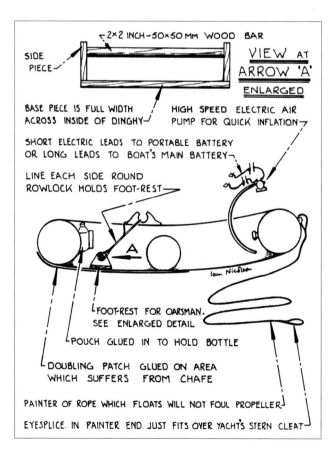